BEHIND THE CLICK. THE UNTOLD STORY OF THE INTERNET

How the Internet Really Works: The Secret Path of Every Click, Stream, and Surf

COPYRIGHT PAGE

BEHIND THE CLICK: The Untold Story of the Internet

For permissions or inquiries, contact:
Cloudoon Press
Fort Worth, Texas, USA
Email: info@cloudoon.com
Website: www.cloudoon.com/press

Publisher Information

Published by Cloudoon Press
Printed in the United States of America
ISBN 979-8-9938117-1-0

Disclaimer

Credits

Author: David Murimi
Editor: Cloudoon Editorial Team
Illustrations & Diagrams: Cloudoon AI Studio
Cover Design: Zola Creative (Cloudoon)

Explaining the Internet — one click at a time.

Second Edition – 2025

Printed in the cloud (well… technically on paper).

DEDICATION

("To everyone who's ever said: Wait, where *is* the Internet?")

CHAPTER 1: THE BIRTH OF THE INTERNET

CHAPTER 2: WHAT HAPPENS WHEN YOU CLICK A LINK

CHAPTER 3: UNDERSTANDING SERVERS & DNS

What is a server, really (and why it's just a fancy computer)

Shared hosting vs. VPS vs. cloud servers

What "the cloud" actually means (hint: still someone's computer)

DNS: The Internet's phone book

Domains, IP addresses, and why names matter

Data centers, cooling, and energy (how your cat videos stay online)

📊 Diagram 3: The Cloud – Data Centers, Servers & You

Bonus Section: The "I Finally Get It" Checklist

Simple questions readers can tick off:

✅ I know what happens when I click a link
✅ I can explain "the cloud" to my grandma
✅ I understand that DNS isn't a disease
✅ I know why websites need servers
✅ I can use "packet" in a sentence without thinking of chips

Glossary: The Internet in Plain English

- **Bandwidth:** How wide your Internet pipe is.
- **Cache:** Your computer's short-term memory for websites.
- **IP Address:** Like your home address, but for your device.
  ...and so on.

CHAPTER 1 – THE BIRTH OF THE INTERNET

Sections

1.1 From "The Cloud" Back to Earth

If you've ever heard someone say, "It's stored in *the cloud*," and imagined your data floating somewhere above the stratosphere—relax, you're not alone. The truth is far less mystical and way more interesting. "The cloud" is really just a bunch of **super-powered computers sitting in warehouses** around the world, quietly humming, blinking, and waiting for you to click something.

But to understand how we got to this invisible-yet-everywhere Internet we take for granted, we have to rewind several decades—back to a time when "going online" meant **physically dialing another machine** over a screeching phone line, and "downloading a file" took longer than cooking dinner.

Before memes, TikTok, and Amazon Prime, there was no Internet—no Google to Google things on. Communication between computers was like teaching toddlers to share toys: everyone had their own system, their own language, and absolutely zero interest in playing nice.

Then came a group of researchers who had an idea so bold it sounded absurd:

> *What if we could connect all computers, everywhere, so they could talk to each other—automatically?*

This wild notion eventually grew into the network that now powers your Netflix binge, your grocery delivery app, and probably the device you're reading this on.

It all started not with tech giants or billion-dollar startups, but with **scientists funded by the U.S. Department of Defense**. Yep—the same folks who build stealth jets also kick-started the system that helps you send cat photos.

In this chapter, you'll meet the awkward teenage years of the Internet: from the government's early experiments, to the first email, to the invention of the protocols that still quietly run under every website today. By the time you're done, the words *router, packet,* and *IP* won't sound like techno-mystical spells anymore—they'll make perfect sense.

1.2 The World Before the Web

Before the Internet, computers were like brilliant introverts — smart, capable, but terrible at socializing. Each one spoke its own private language. If you owned a machine from IBM and your friend had one from DEC, they might as well have lived on different planets. Sharing information meant literally walking across a room (or mailing a floppy disk) and hoping it worked when it arrived.

Back then, "networking" didn't mean adding someone on LinkedIn — it meant physically connecting two machines with a thick cable and praying smoke didn't come out. Files were stored locally, meaning if your computer crashed, your data went down with it. No backups, no cloud, just a lot of coffee and tears.

A World of Islands

Each organization was an island. Universities had their own minicomputers. Government labs had others. Private corporations built internal systems that worked only within their walls. There was no unified way for machines to exchange data outside these silos. Imagine living in a world where every city had its own postal system, none of which delivered outside its borders — that's how computing felt in the 1960s.

In this chaos, a few visionaries at DARPA (Defense Advanced Research Projects Agency) started thinking: *Could we build a "network of networks"?*

That idea would eventually morph into the Internet as we know it.

A Military Problem That Sparked a Digital Revolution

The U.S. military had a communications problem. During the Cold War, they worried about what would happen if a nuclear attack destroyed part of their communication system. Traditional phone networks relied on central switching points — knock one out, and entire regions went dark. The military needed something more resilient — a way for messages to route themselves around damage automatically.

Enter **packet switching** — a concept that would become the Internet's secret sauce. Instead of sending data as one long message (like a phone call), packet switching broke information into small, address-labeled chunks that could travel independently across multiple routes and be reassembled at their destination.

Think of it like sending a pizza one slice at a time through different couriers — each slice knows where it's going, and even if one delivery route fails, the rest can still get there another way.

This concept came from brilliant minds like **Paul Baran** in the U.S. and **Donald Davies** in the U.K., who independently proposed the idea in the early 1960s. Their work inspired DARPA to fund the development of a practical test network: the **ARPANET**.

The Pre-Internet Era at Home

Meanwhile, in the wider world, "computing" meant bulky machines housed in air-conditioned rooms. A single megabyte of storage could cost as much as a small car. The average person interacted with computers through *punch cards* or, later, *terminals* connected to mainframes.

No one had heard of email or websites. Information was physical — stored in filing cabinets, on microfilm, or in your boss's head.

When computers finally crept into homes in the late 1970s and early 1980s, they arrived as single-player islands. You could play games, write documents, maybe dabble in BASIC programming. But connecting to another machine? That was for sci-fi movies.

The Great Connection Gap

What made the Internet revolutionary wasn't computers — it was **communication**. The real magic began when machines learned how to "talk" across distances. The first attempts were crude: users would dial into local bulletin board systems (BBSes) using noisy modems, swapping files and messages over painfully slow phone lines.

But these BBSes were local — like small-town newspapers compared to the global news network that would soon come.

It took government funding, academic curiosity, and a touch of Cold War paranoia to unite all those lonely machines into something bigger.

1.3 How ARPANET Was Born

If the Internet had baby photos, they'd be black-and-white, slightly blurry, and full of guys in short-sleeved shirts staring at beige boxes the size of refrigerators. Those boxes were the first nodes of **ARPANET** — the grandparent of the Internet.

The First Four Computers That Changed Everything

It was 1969. Man had just landed on the Moon, and somewhere in California, a group of researchers were busy connecting four university computers:

1. UCLA (University of California, Los Angeles)
2. SRI (Stanford Research Institute)
3. UCSB (University of California, Santa Barbara)
4. The University of Utah

Together, these formed the first network where computers could actually *send data to each other* electronically — not by hand, not on tape, but through a wired network. It wasn't pretty, but it worked.

The First Message: "LO"

On October 29, 1969, UCLA student **Charley Kline** attempted to log into a remote computer at Stanford. The system promptly crashed after transmitting just two letters: **"L" and "O."**
That was it — the very first Internet message ever sent was literally *"LO."*

If you're into symbolism, that's either a glitchy "LOGIN" or the universe's way of saying **"Hello."**

When the team finally got it working, they'd proven something monumental: two computers hundreds of miles apart could communicate using packet-switched data.

Building the Network Backbone

Once that proof of concept worked, DARPA poured more funding into expanding ARPANET. The backbone grew quickly, connecting universities, military contractors, and research centers across the U.S. Each site had an **Interface Message Processor (IMP)** — a box roughly the size of a washing machine that handled the "plumbing" of data transfer.

Think of IMPs as the Internet's first routers — primitive but revolutionary. They kept the data moving from one node to another, ensuring every packet found its destination even if part of the network went down.

The ARPANET Culture

What's remarkable is that ARPANET wasn't just about technology; it built a **community**. Researchers started using it to share ideas, code, and data long before social media made sharing a pastime.

In 1971, **Ray Tomlinson** sent the first **email** across ARPANET, using the "@" symbol to separate username and host — a tiny decision that still defines digital communication today.

Soon after, **FTP (File Transfer Protocol)** allowed users to exchange files, and **Telnet** made it possible to access remote systems directly. Each of these early tools became a foundation for what we now call the Internet stack.

It's worth noting: there were no memes, no spam, no pop-ups — just pure, academic curiosity. The biggest excitement was logging into a remote mainframe and watching your code actually compile somewhere else.

The 1970s: The Network Expands

By the mid-1970s, ARPANET had grown to over 50 connected institutions. More importantly, the concept of a network of networks was taking hold internationally.

In France, a similar system called **CYCLADES** explored packet-based networking. In the U.K., researchers built **NPL**. Each group had its own protocols, its own quirks — and that fragmentation was about to become a problem.

Different networks couldn't talk to each other. Imagine if iPhones could only call other iPhones and Androids could only call Androids — a frustratingly limited "web" of separate worlds.What the Internet needed was a universal translator.

That's where **TCP/IP** entered the story — the set of rules that made every computer speak the same language.

1.4 Packets, Protocols & Pizza Slices

If ARPANET was the Internet's baby picture, **packet switching** was its heartbeat. Without it, the web would still be a glorified walkie-talkie system — one connection at a time, no multitasking, no memes.

The Genius of Breaking Things Apart

Imagine you order a giant pizza online — extra cheese, please. The restaurant doesn't send the whole pizza intact through one delivery person on a bike. Instead, it cuts it into slices, boxes each one, and sends multiple drivers out using different routes. Even if one driver gets stuck in traffic, the others still arrive. Once all the slices show up, you reassemble the pizza and dig in.

That's **packet switching** in a nutshell.
Every file, video, or message you send online is broken into tiny pieces called packets. Each packet carries:

- A chunk of your data
- The address it's heading to
- Instructions on how to reassemble itself at the destination

This design is what makes the Internet **resilient**. If one route fails, packets automatically find another. Your cat video doesn't stop streaming just because one fiber line somewhere in Tokyo hiccups.

Meet the Protocols: The Internet's Rulebook

Packets are wild children. Without rules, they'd wander around aimlessly. That's where **protocols** come in — the universal agreements that tell computers *how* to communicate.

Think of protocols as traffic laws for data:

- **IP (Internet Protocol):** Handles addressing and delivery. It's the GPS of the Internet.
- **TCP (Transmission Control Protocol):** Ensures all packets arrive safely and in the right order. It's like the friend who checks that everyone made it home from the party.
- **UDP (User Datagram Protocol):** Sends data fast but doesn't care about order — great for live streams and gaming, not for bank transfers.

Together, TCP and IP became the Internet's backbone language — the "universal handshake" that let computers all over the world talk, regardless of brand or location.

Why This Was Revolutionary

Before TCP/IP, connecting networks was like trying to link a toaster to a microwave using dental floss. Everyone used proprietary standards. TCP/IP standardized everything — turning chaos into harmony.

When a computer sends data, TCP breaks it into packets, gives each a label, and tosses them into the IP network. Routers — think of them as postal sorters — forward each packet along the best path. When they reach the destination, TCP reassembles them into a perfect copy of the original file.

It's like magic, except it's math. (And a lot of caffeine-fueled engineering.)

The 1974 Moment

In 1974, computer scientists **Vint Cerf** and **Bob Kahn** published a paper titled *"A Protocol for Packet Network Intercommunication."* It outlined how different networks could talk to each other using TCP/IP — the birth certificate of the modern Internet.

By the early 1980s, the U.S. Department of Defense officially adopted TCP/IP as the standard for ARPANET. That single decision unified dozens of disconnected systems. The Internet as a "network of networks" was born.

Fun Fact: The First Emojis Were Made of Packets

Every emoji you send travels as packets across routers, switches, and servers — hopping thousands of miles in milliseconds. So the next time you text "😂," remember: somewhere, a router in Tokyo just handled that joke.

1.5 The Magic of TCP/IP

If you stripped the Internet down to its bare bones, two things would still be standing: **TCP** and **IP**. Together, they're like Batman and Robin — inseparable, unstoppable, and occasionally misunderstood.

Let's Break It Down

IP (Internet Protocol) is responsible for getting packets from point A to point B. It's the postal service of the Internet — it reads addresses and delivers packages, but it doesn't really care *what's inside* or whether they arrive in perfect order.

TCP (Transmission Control Protocol) is the perfectionist older sibling. It keeps track of which packets made it to the destination, which got lost, and which arrived out of order. If anything's missing, TCP politely asks for a resend.

In other words:

- **IP = "Send this stuff to that address."**
- **TCP = "Make sure all the stuff gets there safely and in order."**

The Three-Way Handshake

When you visit a website, your computer doesn't just start yelling "HELLOOOO!" into the void. It performs a ritual called the **three-way handshake**, which goes like this:

1. **SYN:** Your computer says, "Hey server, can we talk?"
2. **SYN-ACK:** The server replies, "Sure! I hear you loud and clear. Let's do this."
3. **ACK:** Your computer confirms, "Got it. I'm ready. Send the data!"

Only after this digital handshake does actual data — like the contents of your favorite cat video — begin to flow.

It's polite, efficient, and ensures both sides are paying attention before any real exchange happens. Think of it like two people confirming they're on the same phone call before spilling secrets.

Ports: The Internet's Door Numbers

Every device connected to the Internet has an **IP address**, but that's not enough. Imagine an apartment building — the street address (IP) gets you to the building, but **ports** are the individual apartment numbers where specific conversations happen.

- Port **80** is for HTTP (web browsing)
- Port **443** is for HTTPS (secure browsing)
- Port **25** is for email (SMTP)
- Port **22** is for SSH (remote logins)

When you open a web page, your browser is basically saying:

> "Hey, server at 142.251.46.152 (that's Google, by the way), I'd like to talk through door number 443."

The server opens that port, and the conversation begins. It's like ringing the right doorbell on the world's busiest street.

Why TCP/IP Still Rules

The Internet has grown from a few dozen computers to **tens of billions of devices**, yet TCP/IP — invented half a century ago — still runs the show. That's because it was designed to be **scalable and fault-tolerant**.

If one router fails, packets find another route. If one server crashes, the network keeps humming. It's a system that assumes chaos is normal and plans for it.

This resilience is why you can stream Netflix during a thunderstorm, or send an email from an airplane, or join a Zoom call while someone else in your house is hogging bandwidth watching baby goat videos.

A Peek Under the Hood

When you load a website like *truehost.com*, here's what's happening at lightning speed:

1. Your computer sends a DNS query to translate *truehost.com* into an IP address.
2. It then performs the three-way handshake with that IP.
3. TCP breaks your request into packets and numbers them.
4. IP routes those packets across multiple routers, undersea cables, and satellites.
5. The server responds with HTML packets — which your browser reassembles into the web page you see.

All of this happens in a fraction of a second. No one coordinates it manually. It's pure automation — billions of tiny digital agreements working together perfectly (most days).

Fun Fact: TCP/IP Has Been to Space

NASA has used versions of TCP/IP to send data between Earth and the International Space Station. Yes, astronauts check emails, too. (Though latency can make a Zoom call *slightly* awkward.)

Next up: **Section 1.6 – "From Academia to AOL: The Internet Goes Public."**
We'll see how the Internet escaped universities, conquered living rooms, and went from research tool to cultural revolution — complete with dial-up sounds and 1990s nostalgia.

1.6 From Academia to AOL: The Internet Goes Public

By the early 1980s, the Internet was like a cool underground band — loved by scientists and engineers but unknown to the general public. It existed, but only if you worked at a university, a government lab, or a tech company with enough budget to afford a refrigerator-sized computer.

Then something magical happened: ordinary people started getting connected.

The Great Opening: TCP/IP Day

January 1, 1983 — sometimes called **"Flag Day"** in Internet history — marks the moment everything changed. On that day, the entire ARPANET officially switched from its old system (called NCP) to **TCP/IP**, the universal protocol we just explored.

It wasn't glamorous — no fireworks, no press conference — but it unified every network that followed. Suddenly, ARPANET wasn't just a research project; it was the backbone of something far bigger: the **Internet**, with a capital I.

New networks started joining in droves: CSNET (for computer science departments), NSFNET (funded by the National Science Foundation), and countless regional research networks. Each connection multiplied the network's usefulness — the same way adding one more phone makes the whole phone system more valuable.

The network effect had begun.

The Email Revolution

One of the first killer apps of the Internet wasn't social media, or even the Web — it was **email**.

By the mid-1980s, email was exploding across campuses and research centers. For the first time, people could send messages instantly instead of waiting days for snail mail. Scientists were trading research data, professors were sharing jokes, and somewhere, the first spam message was probably being born.

Email made the Internet feel personal. It wasn't just wires and protocols anymore — it was *people talking to people.*

Enter the Nerds (and the Entrepreneurs)

As more folks logged on, private companies began to see opportunity. In 1989, a British scientist named **Tim Berners-Lee**, working at CERN, invented something truly revolutionary: the **World Wide Web**.

The Web wasn't the Internet itself — it was a *service* that ran on top of it, making information accessible through clickable links instead of obscure command-line prompts. Berners-Lee built three things that changed everything:

1. **HTML (HyperText Markup Language):** a way to format documents with links.
2. **HTTP (HyperText Transfer Protocol):** the rules for transmitting those documents.
3. **The First Web Browser and Server:** the tools for viewing and sharing pages.

If TCP/IP was the foundation, **the Web was the front door** — the part regular humans could finally use without an engineering degree.

The 1990s: When Everyone Logged On

Then came the 1990s — a decade of noisy modems, blinking cursors, and the sound that defined an era:

> *Brrrrrr... kssshhhh... beep-beep-beep!*

That was the sound of **dial-up Internet**. To get online, you'd connect your home phone to a modem, dial a number, and wait as your computer screamed in binary until the connection succeeded.

Companies like **AOL (America Online)**, **CompuServe**, and **Prodigy** turned this chaos into a business. They mailed millions of shiny CD-ROMs promising "100 FREE HOURS" of Internet. You'd install their software, type "login," and suddenly be chatting with strangers in New York while sitting in your pajamas in Kansas.

The Internet was no longer a government experiment — it was a social revolution.

The Birth of Browsers and Search

The first mainstream browser, **Mosaic**, launched in 1993. It let users view images and text on the same page — a big deal back then. Mosaic evolved into **Netscape Navigator**, which became wildly popular before being dethroned by **Microsoft's Internet Explorer**.

This era is often called the **Browser Wars** — companies battling for dominance in a suddenly booming world.

And as websites multiplied, something else became desperately needed: a way to *find* things. Enter **search engines** — early attempts like Lycos, AltaVista, and Yahoo paved the way for a small Stanford project that eventually became **Google** in 1998.

Google's simple interface and brilliant algorithm turned the messy sprawl of the early Web into something usable. Overnight, the Internet transformed from a novelty to a necessity.

The Dot-Com Boom (and Bust)

By the late 1990s, everyone wanted in. Startups were popping up faster than pop-ups (which were also invented around then, unfortunately). Investors threw money at anything ending in ".com."

You could order books from **Amazon**, bid on collectibles on **eBay**, and chat with your crush on **ICQ** or **AIM**. Businesses that once needed offices suddenly just needed a website.

Then, in 2000, the **dot-com bubble** burst. Billions of dollars evaporated overnight. Thousands of startups disappeared. But the infrastructure — the servers, cables, and technology — stayed behind. And from that foundation rose the giants we know today: Google, Facebook, YouTube, Amazon, Netflix, and countless others.

A Turning Point

By the early 2000s, broadband replaced dial-up, the Web became interactive (hello, social media!), and the phrase "on the Internet" started to sound silly — because now, we were *living* there.

The academic experiment had evolved into a digital civilization. The Internet was no longer a tool; it was the new global environment — the air modern life breathes.

1.7 The Browser Wars and Search Revolution

If the early Internet was a scrappy garage band, the 1990s browser wars turned it into a stadium tour. Everyone suddenly wanted a piece of the Web — users, corporations, and the world's biggest tech egos.

The Battle for Your Homepage

In 1994, **Netscape Navigator** was the Beyoncé of browsers. It was sleek, easy to use, and made the Web accessible to millions. Clicking links felt like teleportation — one moment you were reading about penguins, the next you were downloading shareware that crashed your computer.

Microsoft, seeing the Internet's potential, didn't plan to sit out. It bundled **Internet Explorer (IE)** with every copy of Windows 95, essentially saying, "Here's your browser — you're welcome." Netscape cried foul, lawsuits flew, and users everywhere discovered the word *antitrust.*

This clash — **Navigator vs Explorer** — wasn't just about software. It was about **who would control access to the Web.** Back then, browsers weren't just tools; they were *gateways* to everything online. Whoever owned that gateway owned the user experience, the ads, the data, and, eventually, the money.

Microsoft's strategy worked. By the early 2000s, Internet Explorer held over 90% of the market. Netscape, like many dot-coms, faded into history (though it lives on in spirit — Mozilla Firefox was built from its ashes).

Meanwhile, in a Dorm Room at Stanford...

While browsers battled it out, two grad students — **Larry Page** and **Sergey Brin** — were annoyed by how hard it was to find things online. Existing search engines ranked pages by how often a keyword appeared, which led to absurd results. If you searched "cat," you got 1000 spammy pages repeating the word "cat" 400 times.

Larry and Sergey had a better idea: **PageRank.** Instead of counting keywords, they counted *links* — reasoning that if lots of websites linked to a page, it was probably important.

Thus, in 1998, **Google** was born.

Google's homepage was shockingly minimalist — just a logo and a search box. No flashing ads, no "Welcome to the Internet!" banners. It loaded fast even on

---

dial-up and gave results that *actually made sense.* It was like finding the librarian who actually knew where the books were.

Within a few years, Google became the front door to the Internet. The Web stopped being something you *surfed* and became something you *searched.*

The Rise of the Platforms

With search came traffic. And with traffic came opportunity. Businesses realized they could build entire empires online if they ranked well on Google. The early 2000s saw the birth of **"Web 2.0"** — a buzzword meaning the Web went from static brochure-style pages to dynamic, user-generated platforms.

Suddenly, the Web wasn't just read-only; it was *write-too.*

- **Wikipedia (2001):** The Internet's collective brain.
- **MySpace (2003):** Where we learned HTML by customizing our profiles.
- **Facebook (2004):** Where we stopped talking to our high-school friends in person.
- **YouTube (2005):** Where cats, tutorials, and viral videos took over bandwidth.

These platforms created a feedback loop: users produced content, content attracted users, and advertisers poured in money. The Internet became *interactive, addictive,* and **profitable**.

Browsers Strike Back

After years of dominance, Internet Explorer began to age poorly — like milk left in the sun. It was slow, insecure, and full of toolbars. (Some PCs had so many that half the screen was buttons.)

Then came **Firefox**, **Safari**, and later **Google Chrome (2008)** — which changed the game again. Chrome was lightning-fast, clean, and synced seamlessly with your Google account. In less than five years, it dethroned IE completely.

Microsoft eventually gave up and built **Edge**, but by then, Chrome had become the default gateway to the digital world. And, in a twist of irony,

---

Google — once the underdog — now controlled both the browser *and* the search engine.

Search 2.0: The Algorithm Era

As billions of pages flooded the Web, Google's algorithms got smarter. Ranking wasn't just about links anymore — it was about *relevance, freshness,* and *trust.*

That's how the modern SEO (Search Engine Optimization) industry was born. Entire professions emerged around figuring out how to make Google happy — writing articles, tweaking titles, and hunting backlinks like digital archaeologists.

If the early Web was a frontier, SEO turned it into a metropolis — structured, optimized, and monetized.

The Invisible Shift to the Cloud

Around the same time, companies like **Amazon Web Services (AWS)**, **Microsoft Azure**, and **Google Cloud** began offering on-demand computing power over the Internet.

Instead of buying physical servers, startups could rent space "in the cloud" — spinning up web apps in minutes instead of months. It was the ultimate evolution of ARPANET's dream: shared computing resources for everyone, anywhere.

When you upload a photo to Instagram or stream Netflix, that data doesn't float in the sky; it lives in gigantic climate-controlled buildings full of servers. The Web's shiny user interfaces sit on top of an enormous hidden layer of steel, silicon, and fiber — **the cloud.**

A World Without Borders

By the 2010s, a kid in Nairobi could start a YouTube channel, a teacher in Mumbai could host classes on Zoom, and a small business in Texas could sell globally through Shopify — all thanks to this layered Internet built over decades.

The Web had become truly **borderless**. But underneath the friendly buttons and cloud emojis, it was still the same thing: TCP/IP packets zipping through routers, exactly as Cerf and Kahn designed them back in the 1970s.

What the Browser Wars Taught Us

1. **Convenience always wins.** Users flock to what's fast, simple, and free.
2. **Monopolies don't last forever.** The underdog of one decade becomes the overlord of the next.
3. **The Web never stops evolving.** From ARPANET to cloud apps, its DNA remains adaptability.

   So next time your browser freezes, remember — it's part of a long, messy, glorious history of human innovation and mild chaos.

1.8 📊 The Internet Family Tree – From ARPANET to the Cloud

HOW THE INTERNET GREW UP: 1969 → TODAY

From 4 Computers to Billions of Devices

1960s — The Roots

- **ARPANET (1969)** - First 4 nodes

UCLA | Stanford | UCSB | Utah

- **Packet Switching** technology invented
- **DARPA Funding** - Cold War motivation

First Network Created

1970s — Standards Era

- **TCP/IP Protocols** developed (1974 - Cerf & Kahn)
- **First Email** (1971) - @ symbol introduced
- **International nets:** NPL (UK), CYCLADES (France)
- Research networks join ARPANET

Foundation Standards Set

1980s — The Expansion

- **"Flag Day" 1983** - ARPANET switches to TCP/IP
- **NSFNET & CSNET** expand to universities
- **"The Internet"** becomes official term
- **Email & FTP** go mainstream in academia

University Network Era

1990s — Web & People

- **Tim Berners-Lee** creates World Wide Web (1989-91)
- **Browsers:** Mosaic → Netscape → IE
- **Search engines** & Dot-Com Boom
- **Dial-up modems** bring web to homes

WWW Goes Public

2000s — Social Web

- **Platforms emerge:** Google, Wikipedia, Facebook, YouTube
- **Broadband** replaces dial-up
- **E-commerce** transforms shopping
- **Mobile devices** begin connecting

Platform Economy Begins

2010s-2020s — Cloud & Beyond

- **Cloud computing:** AWS, Azure, Google Cloud
- **Smartphones** become universal
- **IoT:** billions of connected devices
- **AI & automation** everywhere
- **Streaming** replaces traditional media

Everything Connected

TODAY: THE GLOBAL CLOUD

From 4 computers in 1969 to billions of devices today. The Internet evolved from a research project into the invisible infrastructure powering apps, entertainment, work, and daily life worldwide.

5+ BILLION USERS | GLOBAL DATA CENTERS | EVERYTHING CONNECTED

How the Internet Grew Up: 1969 → Today

Layer 1 – 1960s – The Roots

ARPANET (1969)
- 4 nodes
- DARPA Funding + Cold War Motivation

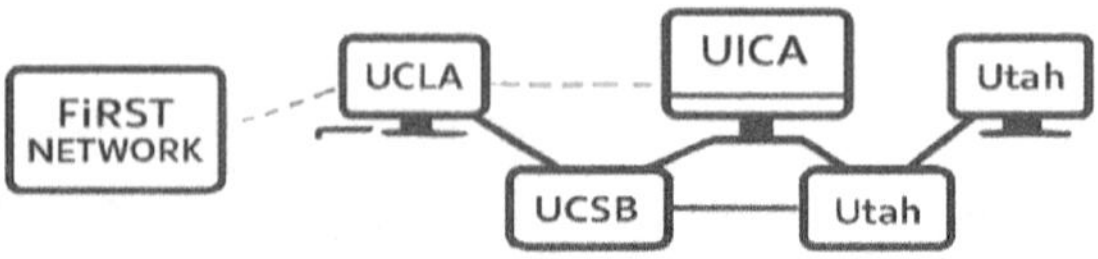

Layer 2 – 1970s – The Standards Era

TCP + IP Protocols developed (1974 – Cerf & Kahn
- First Email (@) #symbol introduced
- Networks like NPL (UK) and CYCLADES (France) appear

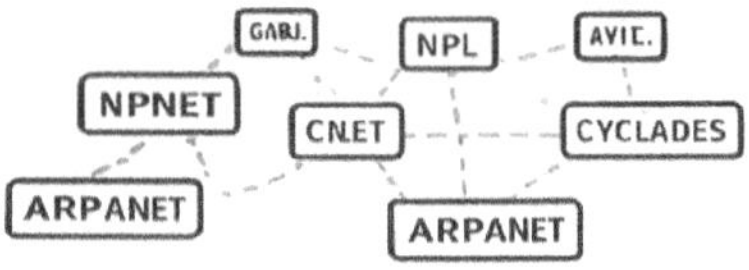

Layer 3 – 1980s – The Expansion

"Flag Day" 1983 ARPANET switches to TCP/IP
- NSFNET+ CSNET expand to universities
- Term: *The Internet*' becomes official

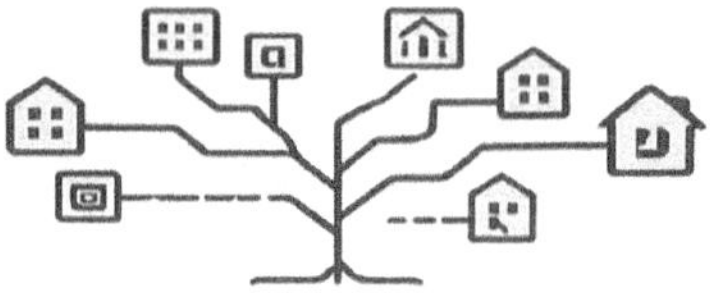

Layer 4 – 1990s – Vhe Web and People

Tim Berners-Lee creates World Wide Web (1889–91)
- Browsers: Mosaic → Netscape entenret
- Search Engines and Dot-Com Boom
- Dial-Up Modems → "WWW Era"

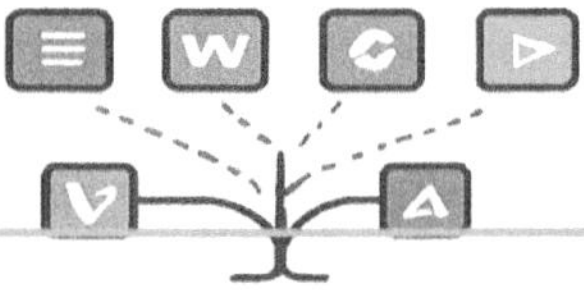

Layer 5 – 2000s – Platforms and Social Web

Google, Wikipedia, Facebook (YouTube
- Broadband replaces Dial-Up
- E-Commerce + Mobile Devices appear
- AI & Automation connect billions of devices

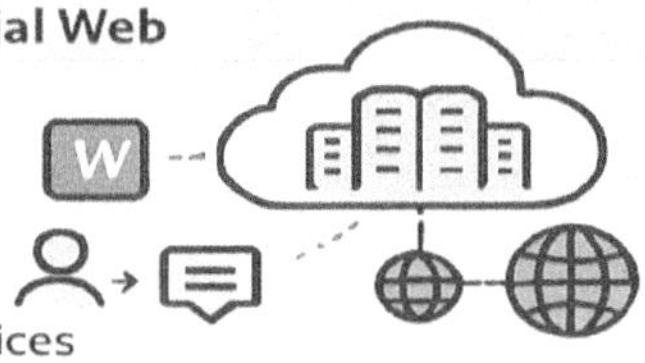

From four computers in 1969 to billions of devices today — and it all still runs on the same basic idea: send tiny packets of data anywhere, anytime.

1.9 Summary & Bridge to Chapter 2 – From History to Action

Take a deep breath — you've just traveled through fifty years of Internet evolution in one chapter. 🎉

From dusty research labs and clunky mainframes to sleek smartphones and cloud servers, the story of the Internet isn't really about technology — it's about **people finding better ways to connect**. Every breakthrough we discussed came from someone asking, "What if we could talk faster, share more, or make this easier?"

Let's do a quick recap before we move on to the fun part — *what actually happens when you click something.*

Key Takeaways from Chapter 1

1. The Internet started as a defense experiment.
ARPANET was designed to keep communication alive, even if part of the network failed. The Cold War's paranoia accidentally gave birth to your TikTok habit.

2. Packet switching was the game-changer.
By slicing information into packets, data could travel anywhere, re-route automatically, and reassemble perfectly. That's why the Internet survives broken cables, storms, and the occasional squirrel chewing a fiber line.

3. TCP/IP made everyone speak the same language.
Once these universal protocols were adopted, computers across universities and continents could finally talk — the moment the "network of networks" truly began.

4. The World Wide Web made it usable.
Tim Berners-Lee's simple idea of linking documents turned the Internet from a geeky tool into a global playground.

5. Browsers, search engines, and the cloud democratized access.
From Netscape to Google Chrome, from dial-up tones to Netflix streams, the Internet became faster, friendlier, and (sometimes) a little too addictive. Behind it all, those same TCP/IP packets are still doing the heavy lifting.

From Then to Now

When people say "it's stored in the cloud," you now know the truth: it's just data traveling through a vast maze of cables, routers, and servers — all built on principles invented before disco died.

Every email, photo, and tweet you send is basically a digital postcard hopping across networks at near-light speed, following rules first scribbled down by researchers in the 1970s.

So the next time you tap a link, pause for a second and appreciate that you're commanding one of the most complex systems ever built — with a single fingertip.

Up Next – Chapter 2: What Happens When You Click a Link

Now that you know *where* the Internet came from, it's time to peek *inside* it.

In Chapter 2, we'll trace one simple action — clicking a link — and follow your request as it travels through your Wi-Fi router, your Internet Service Provider, across DNS lookups, into distant data centers, and back again as a perfectly rendered web page.

It's the Internet's version of a road trip movie — except the cars are data packets and the roads are invisible. Buckle up; your browser's about to hit the highway. □□

✅ **End of Chapter 1 — "The Birth of the Internet."**

CHAPTER 2 – WHAT HAPPENS WHEN YOU CLICK A LINK

Sections:

2.1 The Click Heard Around the World (Intro Hook)
2.2 Your Browser Makes a Request
2.3 DNS: The Internet's Phone Book
2.4 The Trip Through Your ISP
2.5 Servers, HTTP, and the Response
2.6 Rendering the Page: How Your Screen Comes Alive
2.7 Common Errors (and What They Really Mean)
2.8 📊 *Diagram Placeholder – "Anatomy of a Web Request"*
2.9 Summary & Bridge to Chapter 3

2.1 The Click Heard Around the World

Let's imagine it's a normal Tuesday morning. You open your laptop, sip your coffee, and click a link — maybe to check your email, watch a video, or see what that one friend posted at 3 a.m.

Instantly, the page appears on your screen. It feels like magic. But what really happened under the hood in that split second?

Spoiler: your "click" just triggered a digital relay race involving **millions of lines of code, thousands of machines, and data flying through fiber-optic cables at nearly the speed of light.**

Seriously — all that for a cat meme.

It All Starts With You

When you click a link (say, https://www.truehost.com), your browser — Chrome, Safari, Edge, or whichever you prefer — takes that as a command: *"Find this website, grab whatever it's got, and show it to the user."*

But your computer doesn't actually know what "truehost.com" means yet. It needs to find its **numerical address** — like looking up someone's phone number before calling them. That's where the **Domain Name System (DNS)** comes in, which we'll explore in detail in a moment.

First, though, your computer packages that request into little data bundles, called **packets**, marked with an instruction that essentially says,

> "Hey Internet, can you please deliver this to whoever runs truehost.com?"

And then — off they go, zipping out of your Wi-Fi router and into the vast global network.

A Worldwide Game of Hot Potato

From there, those packets jump across routers, modems, switches, and ISPs — each passing the data to the next closest network until it reaches its destination. Every router reads the packet's "To:" label (the IP address) and decides, "Where's the fastest way to forward this next?"

It's like a digital FedEx system, except instead of delivery vans, you've got photons racing through glass fibers thinner than a human hair.

And it all happens so fast that before you can even say "loading," your request is halfway across the world.

The Moment Before the Miracle

When your packets finally find the web server hosting the site you asked for, that server sends back an answer — the web page, images, and code that your browser will piece together into something beautiful (or sometimes hideous).

The entire round trip — from click to page — usually takes less than one second. Faster than you can blink.

But that single second hides an incredible ballet of machines cooperating perfectly across continents.

2.2 Your Browser Makes a Request

If the Internet were a play, your **browser** would be the lead actor — performing your requests, memorizing lines, and occasionally forgetting its cues when too many tabs are open.

When you click a link, the browser springs into action faster than a caffeinated intern on deadline. Its mission: figure out *where* to send your request, *how* to send it, and *what* to ask for.

Step 1: Parsing the Link

Let's say you click this:

https://www.truehost.com/contact-us

To a human, that just means "take me to the Contact page."
To your browser, it's a coded message containing everything it needs:

Part	Meaning
https://	The protocol — tells the browser *how* to talk to the server (securely, in this case).
www.truehost.com	The domain name — the friendly alias for the site's IP address.
/contact-us	The path — the specific file or page to fetch.

It's like an envelope with the address, delivery method, and instructions on which room to knock on.

Step 2: Checking the Cache

Before your browser sends the request out to the wild Internet, it checks its own **cache** — a short-term memory of sites you've recently visited.

If you've opened the same page before and nothing's changed, your browser can pull it straight from local storage instead of downloading it again.

This is why sites you revisit often seem to "load faster." It's also why your ex's old Instagram page might still show up even after they've deleted it — the browser is *helpfully* clinging to the past.

Step 3: DNS Lookup — Finding the Address

If the page isn't cached, the browser needs to find out **where** "truehost.com" lives — its numerical IP address.

Think of DNS (Domain Name System) as the Internet's giant phone book. It converts human-friendly names (like truehost.com) into machine-friendly numbers (like 104.21.12.143).

Your browser asks your operating system,

> "Hey, do we already know the IP for this domain?"
> If the OS says no, the query goes to your **DNS resolver** — usually run by your Internet Service Provider (ISP) or a public DNS like Google's (8.8.8.8).

Once it gets the answer, it stores it locally for next time. Efficient and lazy — the Internet's favorite combo.

Step 4: Building the HTTP Request

Now that it knows where to send the data, your browser constructs a formal request message. This isn't casual small talk — it's written in **HTTP (HyperText Transfer Protocol)**, the language of the Web.

Here's what that message roughly looks like under the hood:

```
GET /contact-us HTTP/1.1

Host: www.truehost.com

User-Agent: Chrome/121.0

Accept: text/html
```

Let's decode that:

- **GET** tells the server, "Please send me this resource." (There are other commands like POST, PUT, DELETE — but GET is the classic.)
- **Host** specifies which domain you're contacting.
- **User-Agent** identifies your browser and version (so websites can tailor content).
- **Accept** declares what kind of data you can handle — HTML, JSON, etc.

It's like writing a polite letter that says,

> "Dear Server, I'd love a copy of your Contact page. I promise not to break anything. Sincerely, Chrome on Windows."

Step 5: Sending the Request

Finally, your browser sends this carefully written HTTP request out through your computer's network interface, into your Wi-Fi router, and out to your Internet Service Provider.

But remember — the Internet doesn't move one big message at a time. The browser chops it into packets, each labeled with:

- Source IP (your device)
- Destination IP (the server)
- Port number (which "door" the request is entering)
- Sequence number (so they can be reassembled later)

Each packet then begins its journey across routers, fiber cables, and switches — a digital road trip across the globe.

Step 6: Waiting... Patiently (Sort Of)

While all this happens, your browser twiddles its thumbs — metaphorically. In reality, it's opening a **TCP connection** (using that three-way handshake from Chapter 1), making sure both sides are ready to communicate securely.

If you're visiting an HTTPS site, there's an extra layer of security here: **TLS (Transport Layer Security)**.
It's like whispering instead of shouting in public. The browser and server exchange encryption keys to ensure no one can eavesdrop on your conversation.

This is what gives you that comforting little padlock icon in the address bar — your guarantee that no hacker or nosy barista can read what's being transmitted.

Step 7: The Request Leaves the Building

Once everything checks out, your browser finally sends the request into the wild. From that moment on, your data starts bouncing between routers like a digital game of pinball.

Within milliseconds, it's out of your home, through your ISP, and racing across the Internet backbone — a web of high-speed cables and massive data centers linking continents.

Your part is done. The baton has been passed.
Now it's up to the Internet's invisible infrastructure to deliver your tiny message to the right place — and, more importantly, bring back the answer.

2.3 DNS: The Internet's Phone Book

Let's be honest — nobody wants to remember long strings of numbers like 142.251.46.142.
That's Google's IP address, by the way. (Yes, it really is a phone number for computers.)

Humans are great at remembering names like *google.com* or *truehost.com*, but computers only understand numbers.
So the Internet, being a clever multitasker, invented a translator: the **Domain Name System**, or DNS.

Why DNS Exists

Imagine you want to call your friend, but instead of dialing their name in your contacts list, you had to memorize their 10-digit number every time. Brutal, right?

That's how the Internet used to be — computers connecting to other computers by typing in their numerical "phone numbers" (IP addresses). It was like trying to remember everyone's number in the world, but worse, because websites can move, change addresses, or live on multiple servers.

DNS fixed all that. It's the Internet's **global contact book**, mapping readable names like *truehost.com* to their numerical addresses. Every time you visit a site, your browser quietly asks DNS,

> "Hey, what's the number for this name?"

DNS answers in milliseconds, and off your packets go.

How the DNS System Works

Think of DNS as a **layered hierarchy** — kind of like how you'd look up someone in a phone book (if you're old enough to remember those).
Here's how your computer finds the IP address for *www.truehost.com*:

Step 1: The Local Cache

First, your browser checks if it already knows the answer from a previous visit. If it does — boom, instant access.
If not, it asks your computer's operating system (Windows, macOS, Linux). Still no luck? Time to climb the ladder.

Step 2: The Resolver (Your ISP or a Public Service)

The next stop is your **DNS resolver** — usually provided by your Internet Service Provider.
This resolver is like a personal assistant who goes out and does the lookup on your behalf. It doesn't know everything, but it knows who to ask.

Step 3: The Root Servers

If your resolver doesn't have the address, it contacts one of the **13 root server clusters** that sit at the top of the Internet's addressing system.
These aren't single machines — they're global networks run by organizations like Verisign and ICANN, spread across hundreds of locations for redundancy.

The root servers don't know every website, but they *do* know where to find the next step — the **Top-Level Domain (TLD) servers**.

Step 4: The TLD Servers

The TLD servers handle extensions like *.com*, *.net*, *.org*, *.us*, *.ng*, etc.

So if you're looking for *truehost.com*, the root server says, "Ask the .com directory over there."
The TLD server then points your resolver to the right **authoritative name server** for that specific domain.

Step 5: The Authoritative Name Server

Finally, your resolver reaches the authoritative name server for *truehost.com* — usually managed by the company that owns or hosts the domain.
This server is the final boss in the DNS chain. It holds the definitive record of where that domain currently lives — the IP address you're looking for.

Your resolver takes that answer (say, 104.21.12.143), caches it for a while, and passes it back to your browser.
Your browser then says, "Aha! Now I know where to send my request."

All of this happens in a blink — often faster than it takes to say "DNS lookup."

DNS Records: The Internet's Address Cards

Inside those authoritative servers are little digital index cards called **DNS records.**
They come in different types, each serving a purpose:

Record Type	What It Does
A Record	Maps a domain to an IPv4 address (like 104.21.12.143).
AAAA Record	Maps to an IPv6 address (a longer, newer format).
CNAME	A nickname that points one domain to another (e.g., www → main domain).
MX Record	Tells email where to go (Mail Exchange).
TXT Record	Holds extra info — often for verification or security (like SPF, DKIM).

These records are what make everything from your Gmail inbox to your e-commerce checkout function correctly.

TTL: How Long DNS Remembers

Every DNS record has a **TTL (Time To Live)** — how long a resolver should keep the information before asking again.
If TTL is set to one hour, your ISP will remember the IP for that domain for exactly 60 minutes before re-checking.

Low TTL = fresher data but more lookups.
High TTL = faster access but stale info if something changes.

If you've ever switched web hosts and your site took "a few hours to update," that's TTL in action.

DNS Caching: The Internet's Secret Speed Hack

To speed things up, DNS data is cached at multiple levels — your browser, your OS, your router, and even your ISP.
It's like a digital whisper chain: once someone learns an address, everyone remembers it for a while.

That's why when a website moves to a new server, some people see the new version instantly, while others keep hitting the old one for hours.
The Internet, it turns out, has a slightly delayed memory.

When DNS Goes Wrong

A mistyped domain? You end up on a "404" page.
A hacked DNS record? You could land on a fake website designed to steal your info — a **phishing attack**.
That's why DNS security matters.

Modern systems use **DNSSEC (Domain Name System Security Extensions)** — digital signatures that ensure responses haven't been tampered with.
It's like having caller ID for your Internet lookups.

Fun Fact: DNS Has Its Own Celebrities

Public DNS providers have catchy, easy-to-remember IPs:

- Google: 8.8.8.8 and 8.8.4.4
- Cloudflare: 1.1.1.1 (they claim it's the fastest)
- OpenDNS: 208.67.222.222

They all do the same basic thing — translate names to numbers — but some do it faster, more securely, or with better privacy.

In short, DNS is the Internet's unseen concierge — the system that makes sure every digital letter you send ends up in the right mailbox. It doesn't get headlines, but without it, you'd be staring at a blank screen, typing numbers like it's 1973.

2.4 The Trip Through Your ISP

Now that your browser has an IP address and a neatly packaged request, it's ready for a field trip — through the wild, tangled highways of the Internet. But before your packets can mingle with the world, they first have to pass through the local gatekeeper: your **Internet Service Provider**, or **ISP**.

If the Internet were an airport, your ISP would be the check-in counter, baggage screener, and air traffic controller rolled into one.

Meet Your Home Network Crew

Let's start at home. When you connect to Wi-Fi, your device is actually talking to a small chain of helpers:

1. **Your Device (Laptop/Phone):** The chatterbox who wants to go places.
2. **Your Router:** The traffic cop managing multiple devices (laptop, TV, smart fridge, that one game console you regret buying).
3. **Your Modem:** The translator that speaks your ISP's language — coaxial cable, fiber, or DSL.

Your router gives each device in your home a *private* IP address (usually something like 192.168.0.12). These aren't visible to the outside world — they're like your family's nicknames that only work inside your house.

When your laptop sends a request to www.truehost.com, your router intercepts it and says, "Okay, I'll make sure this gets out safely."

The modem then converts the digital signal into whatever physical format your ISP expects — flashes of light for fiber, radio waves for wireless, or good old-fashioned voltage changes for copper cables.

Step 1: Your ISP's Local Network

Once the data leaves your modem, it enters your ISP's local network — basically a neighborhood hub that connects thousands of customers like you.

Every ISP has local routers that gather traffic, sort it, and send it toward its destination using — you guessed it — **packets.**

At this stage, your packets are just one among millions flowing through the same digital bloodstream. Your neighbor might be streaming Netflix, someone else is on a Zoom call, and you're trying to load a cat meme. All those packets share the same physical lines but are kept perfectly separate by tagging and timing.

Your ISP's routers use sophisticated algorithms to decide where to send each packet next. They don't "see" your data's contents (especially if it's HTTPS encrypted), but they do see the destination IP and make real-time routing decisions.

Step 2: Peering and Transit – The Internet's Intersections

Here's where things get global.

Your ISP isn't connected directly to every website in the world — that would be impossible. Instead, ISPs form **peering agreements** with one another and with larger network operators to exchange traffic.

Think of it like highways connecting different cities. Your local ISP has an on-ramp to major Internet backbones — enormous, high-capacity cables that carry traffic between countries and continents.

When your packets hit these backbones, they can travel thousands of miles in milliseconds. Under the Atlantic Ocean alone lie dozens of fiber-optic cables, each thinner than a garden hose but capable of transmitting hundreds of terabits per second — that's billions of YouTube videos at once.

These backbone providers include names you've probably never heard — Level 3 (now Lumen), Cogent, Telia, GTT — but without them, there'd be no "global" in "global network."

Step 3: Hops, Routers, and the Digital Road Trip

Every router your packet passes through is called a **hop**. You can actually trace this journey yourself using a command called **traceroute** (on Windows: `tracert truehost.com`, on macOS/Linux: `traceroute truehost.com`).

It will show you every stop along the way — sometimes a handful, sometimes 20 or more, depending on where the server lives.

For example, if you're in Dallas visiting a site hosted in London, your packets might go:

Dallas → Chicago → New York → London → Data Center

Each hop is a router reading your packet's header, deciding where next to send it, and handing it off like a relay baton.
Routers communicate using protocols like **BGP (Border Gateway Protocol)** — the Internet's GPS system that keeps track of which networks know how to reach which addresses.

When something goes wrong — like a fiber cut in Tokyo — BGP automatically reroutes traffic, sometimes within seconds, to avoid downtime. That's why you rarely notice when parts of the Internet go down.

Step 4: Quality, Speed, and Bottlenecks

Not all routes are created equal. The Internet's "roads" vary in quality, speed, and congestion.

- **Latency** measures how long it takes for your packets to reach their destination and come back (ping time).
- **Bandwidth** measures how much data can move at once.
- **Jitter** is how much that latency fluctuates. (Gamers hate jitter.)

Your ISP and backbone providers constantly balance these factors, rerouting traffic to keep things smooth. But during peak hours — like Friday nights — some paths get jammed. That's when YouTube might buffer, and you start glaring at your router like it owes you money.

Step 5: Crossing Borders and Security Checks

When your packets travel internationally, they often cross through Internet exchange points (IXPs) — giant data centers where networks from different countries meet and trade traffic.

These IXPs are the beating hearts of the Internet, moving petabytes of data every second. Some famous ones include:

- **LINX** in London
- **DE-CIX** in Frankfurt
- **AMS-IX** in Amsterdam
- **Equinix** hubs across the U.S.

Despite the massive scale, your privacy is largely intact — especially on encrypted connections. Governments and ISPs can see *where* you're connecting, but not *what* you're doing (thank you, HTTPS).

Step 6: Arrival at the Destination Network

Eventually, your packet reaches the **destination network** — the one that owns the IP address you were seeking. This could be:

- A small web host running a few servers,
- A massive cloud provider like Amazon Web Services, Google Cloud, or Microsoft Azure,
- Or even a distributed content network like Cloudflare or Akamai.

At that point, your request leaves the ISP world and enters the data center's realm — where servers, load balancers, and firewalls decide what happens next.

So Where Is "The Internet," Really?

By now, you might have noticed something interesting: there's no single "place" called *the Internet.*
It's a web of connections, contracts, and cables — millions of networks voluntarily cooperating through shared rules (like TCP/IP and BGP).

Your ISP is your personal gateway into that web. Without it, your computer would be a lonely island — full of potential, but talking to no one.

So next time your connection drops, take a breath before yelling at your modem. Remember, your data is traveling across oceans, through lasers, bouncing between routers in multiple countries — all in less than a second.
A little lag is basically a miracle.

2.5 Servers, HTTP, and the Response

Your packet has had quite a journey — through routers, ISPs, undersea cables, and Internet exchanges. Now, it finally reaches the destination: a **server**, sitting quietly in a data center somewhere, blinking away under fluorescent lights and freezing air conditioning.

And just like that, the story flips. Your computer stops being the one asking questions. The server takes the stage to reply.

So What's a Server, Really?

A **server** is just a computer — not that different from yours, but built for endurance, speed, and scale.
Where your laptop has one user, a server can handle thousands (or millions) of requests every minute. It doesn't nap, play Spotify, or accidentally close all its tabs. It's a 24/7 machine whose only job is to **serve** information when asked.

There are many kinds of servers:

- **Web servers** deliver websites (Apache, Nginx, IIS).
- **Database servers** store and manage information.
- **Application servers** handle logic, like login systems or online payments.
- **Mail servers** manage your inbox.

All these work together, often in the same data center, to make a website load in the blink of an eye.

Step 1: The Server Receives Your Request

When your browser's HTTP request arrives at the destination IP address, it lands at the **front door** — a port, typically port **80** (for HTTP) or **443** (for HTTPS).

A web server program (like Nginx or Apache) is listening there 24/7, ready to say,

> "Okay, what do you want?"

It reads your request line — for example:

```
GET /about.html HTTP/1.1

Host: www.truehost.com

User-Agent: Chrome/121.0
```

The server checks:

- Is this domain actually hosted here?
- Does /contact-us exist?
- Does the user have permission to access it?

If everything checks out, the server gets to work finding the content.

Step 2: The Server Talks to Its Friends

Most modern websites aren't just a bunch of static HTML files sitting in a folder. They're dynamic — built on the fly every time you visit.

For example, when you load a blog or an online store:

1. The web server passes your request to an **application layer** (like PHP, Python, Node.js, or Ruby).
2. That layer talks to a **database server** (like MySQL or PostgreSQL).
3. The database sends back the requested data — say, the text of a blog post or the list of products.
4. The application assembles that data into an HTML page.
5. The web server wraps it up and sends it back your way.

All of that happens in milliseconds — thousands of queries, responses, and logic checks just to show you a web page with a few words and images.

It's like a restaurant where the waiter (the web server) takes your order, the kitchen (the database) prepares the meal, and the server (the application) plates it nicely before bringing it to your table. □□

Step 3: Sending the Response

Once the server has prepared the content, it wraps it in an **HTTP response** — the flip side of your original request.

Here's what a simple one looks like:

```
HTTP/1.1 200 OK

Content-Type: text/html; charset=UTF-8

Content-Length: 1256

Date: Tue, 03 Nov 2025 15:12:00 GMT

Server: nginx/1.25.1
```

Followed by the HTML code for the page.

Let's break that down:

- **200 OK:** Success! Everything worked fine.
- **Content-Type:** Tells the browser what kind of data this is (HTML, JSON, image, etc.).
- **Content-Length:** How big the payload is.
- **Date & Server:** Metadata — useful for debugging and bragging rights.

The HTML content follows right after those headers — the actual code your browser will later turn into a beautiful web page (or a chaotic one, depending on the designer).

Step 4: Error Codes – When Things Go Wrong

Of course, things don't always go smoothly. Servers also return status codes when something goes wrong — the polite way of saying, "Oops."

Code	Meaning	Typical Cause
404 Not Found	The page doesn't exist.	Broken link, typo, deleted file.
403 Forbidden	You're not allowed in.	Permissions issue, restricted access.
500 Internal Server Error	Something broke.	Server misconfiguration, bad code.
301/302 Redirect	The page moved elsewhere.	Common during site redesigns.
503 Service Unavailable	Server overloaded or under maintenance.	Too many users, server meltdown.

The next time you see a 404 page, remember: your request reached the server, but the server shrugged and said, "Sorry, I don't have that."

Step 5: The Data Goes Back Home

Once the response is ready, the server begins sending packets back along the same route — routers, ISPs, undersea cables, back to your modem.

But here's the twist: it doesn't always take the exact same path.
Internet routing is dynamic — packets can find the fastest available way back, like GPS recalculating mid-drive.

Your computer reassembles those packets, TCP ensures none were lost or out of order, and the browser starts decoding the HTML line by line.

As soon as enough data arrives, the browser begins **rendering** — building the visible page even before the rest finishes loading. That's why you often see text before images or a "skeleton" of the site before everything appears fully styled.

The Modern Twist: CDNs and the Cloud

Most popular websites today don't rely on one server. Instead, they use **Content Delivery Networks (CDNs)** — distributed networks of servers around the globe that cache content close to users.

So if you're in Kenya visiting a U.S.-based site, you might actually be fetching the data from a local node in Nairobi — same content, closer to home, faster response.

Companies like **Cloudflare**, **Akamai**, and **Amazon CloudFront** handle billions of these transactions daily. They also filter out bad traffic (like bots and hackers) before it even reaches the main servers.

In short, your request rarely travels all the way across the world anymore — it just looks like it does. The cloud keeps copies of popular data scattered everywhere, ensuring speed, redundancy, and resilience.

And Just Like That…

In less than a second, your tiny packet completed a round trip that would make Phileas Fogg jealous — from your home to a server possibly halfway around the planet and back, all to show you one page.That's the real magic of the Internet: billions of machines, none in charge, all cooperating perfectly (well, mostly) to deliver what you want, when you want it.

2.6 Rendering the Page: How Your Screen Comes Alive

You've clicked the link.
Your packets traveled halfway around the planet and back.
The server replied.

Now your browser rolls up its sleeves and says,

> "Okay, let's make this thing beautiful."

Rendering a webpage is like putting together IKEA furniture: you've got the instructions (HTML), the paint (CSS), the moving parts (JavaScript), and somehow it all has to fit together without collapsing.

Step 1: The HTML Arrives — The Blueprint

The first thing your browser gets from the server is usually **HTML** — HyperText Markup Language. It's not a programming language, it's more like a *blueprint* for your page's structure.

Imagine a skeleton:

- The **head** section includes behind-the-scenes info: title, metadata, links to stylesheets, scripts, and SEO details.
- The **body** contains the visible parts — headings, paragraphs, buttons, forms, images, videos, and all that good stuff.

Here's a tiny example:

```
<!DOCTYPE html>
<html>
<head>
  <title>About Us - Example Co.</title>
</head>
<body>
  <h1>Welcome to Example Co.</h1>
  <p>We make the Internet a friendlier place.</p>
</body>
</html>
```

Your browser doesn't "see" the page like a human — it reads the tags (<h1>, <p>, <img>) and builds a model of what should appear. This invisible structure is called the **DOM** (Document Object Model).

Think of it as the page's family tree — every element (like a paragraph or image) is a branch the browser can manipulate later.

Step 2: Styling It Up with CSS

Next, the browser grabs the **CSS (Cascading Style Sheets)** — the part that makes websites look good (or, in some cases, look like a design intern got too creative).

CSS controls everything visual:

- Fonts, colors, spacing, layout, animations.
- Responsive design — how things rearrange on phones vs desktops.

If HTML is the skeleton, CSS is the outfit, makeup, and lighting combined.

Example:

```
h1 {
  color: #FFA500;
  font-size: 36px;
  text-align: center;
}
```

Now your boring <h1> becomes a bright orange headline — fittingly, Truehost orange (#FFA500 😎).

Your browser merges the HTML structure with the CSS rules to create a **render tree**, mapping out what each element should look like and where it belongs on the screen.

Step 3: JavaScript Joins the Party

Then comes **JavaScript** — the hyperactive cousin of HTML and CSS. It makes pages *do* things.

Whenever you see buttons that expand, sliders that move, or live content that updates without reloading — that's JavaScript at work.

For example:

```
document.getElementById("greet").innerHTML = "Hello, Internet!";
```

This single line of code could instantly change text on your page, fetch live data, or play an animation.
Modern web apps (like Gmail, YouTube, or TikTok) are mostly powered by JavaScript frameworks like **React**, **Vue**, or **Angular**, which make websites feel more like full-fledged apps.

When your browser runs JavaScript, it uses a built-in engine (like Chrome's *V8*) that executes the code line-by-line, updating the DOM as needed.
It's like a stage crew constantly moving props while the play is already running.

Step 4: The Render Pipeline – The Browser's Assembly Line

Let's recap what happens inside your browser after the data arrives:

1. **Parsing:** It reads the HTML and builds the DOM.
2. **Style Calculation:** It processes CSS and applies styles.
3. **Layout (Reflow):** It figures out where everything goes on the screen.
4. **Painting:** It fills in colors, text, and images pixel by pixel.
5. **Compositing:** It layers all those painted sections into the final image you see.

This process happens dozens of times per second while you scroll, click, or resize the window — all while keeping animations smooth and videos in sync.

If it sounds intense, that's because it is. Your browser is basically performing live theater with thousands of moving parts, every second.

Step 5: Loading Additional Resources

As the page builds, the browser keeps fetching more resources in the background:

- **Images and icons** (.jpg, .png, .svg)
- **Stylesheets** (.css)
- **Scripts** (.js)
- **Fonts** from Google or other CDNs
- **APIs** — data pulled from other servers

Each of these creates a new network request, sometimes to completely different domains.
That's why your browser's "loading" spinner can last a few seconds — it's juggling dozens of simultaneous downloads, verifying certificates, and decompressing files.

The goal is to make it feel seamless, even though the browser might be having a mild panic attack behind the scenes.

Step 6: Security and Sandboxing

Modern browsers don't just render pages — they also **protect** you.
Each tab runs in its own **sandboxed process**, meaning if one page crashes or gets infected, it can't take the rest down with it.

JavaScript is also tightly controlled; it can't access your personal files or spy on your webcam (unless you allow it).
Every website runs in a virtual bubble, talking to the outside world only through approved APIs.
That's why you sometimes see messages like "This site wants to access your location" — the browser is asking on its behalf.

Step 7: The Page Comes Alive

Finally, everything assembles. The layout stabilizes, images pop in, fonts smooth out, and animations begin to play.
At this point, you see a fully rendered, interactive website.

You might think, "Nice, it's loaded," but behind the curtain, your browser is still working — keeping scripts running, listening for clicks, checking for updates, caching images, and preloading content it predicts you'll need next.

Browsers today are basically miniature operating systems, each managing hundreds of tiny processes at once.
And you thought clicking a link was simple.

A Fun Way to Think About It

Here's an analogy:

- **HTML** = The house frame.
- **CSS** = The paint, furniture, and decor.
- **JavaScript** = The electricity, plumbing, and smart devices that make things interactive.
- **Browser** = The construction crew that builds it all in seconds every time you knock on the door.

Every web page you visit is built like this — on the fly, in milliseconds, billions of times per day. That's the invisible magic powering everything from Google search results to your favorite online pizza order.

2.7 Common Errors (and What They Really Mean in depth)

f the Internet were a restaurant, most of the time your order comes out perfectly — hot, fast, and exactly what you asked for.
But sometimes, the waiter (your browser) brings you an empty plate and says, "Sorry, the kitchen had *issues.*"

That's what error codes are — polite ways of telling you something somewhere n the network had a mini nervous breakdown.

Let's lift the curtain and translate these cryptic messages into plain English.

404: Not Found — "We Swear It Was Here Yesterday"

This is the Internet's most famous error, practically a celebrity.
404 Not Found means your browser reached the right server… but the specific page you wanted doesn't exist.

Maybe it was deleted. Maybe it was moved. Maybe you just mistyped page.html instead of page.htm.

To visualize this: you walk into a store asking for "glitter glue, aisle seven," and the clerk shrugs, "We *used* to carry that."

Some sites turn 404s into art — custom pages with jokes or search bars. Others just slap a boring error message. But deep down, all 404s whisper the same thing: *"You're lost, friend."*

403: Forbidden — "You Shall Not Pass"

This one sounds dramatic because it is.
A **403 Forbidden** error means the server understood your request but refuses to fulfill it.

It's not a typo. You found the right place; you just don't have permission to enter.

Common causes:

- Trying to access private files or admin areas.
- The website's permissions are misconfigured.
- You're sneaking into restricted content (shame on you 😅).

Basically, the server is Gandalf standing at the bridge shouting, *"You shall not* pass!"

500: Internal Server Error — "The Kitchen Caught Fire"

500 Internal Server Error is the Internet's way of saying, "Something broke, and we have no idea what."

This is the most dreaded one for developers because it's so vague. It could mean:

- The site's code crashed mid-request.
- The database is down.
- A script forgot how to behave.

From your side, all you can do is refresh and hope someone in the data center is putting out the metaphorical fire.
It's not your fault. It's theirs.

502: Bad Gateway — "Middleman Drama"

Sometimes your request has to pass through intermediaries — load balancers, proxies, or content delivery networks (CDNs).
A **502 Bad Gateway** means one of those middlemen couldn't talk properly to the next link in the chain.

Imagine ordering pizza through five delivery apps chained together. Somewhere along the line, someone dropped the box, and everyone's now pointing fingers.

It's usually temporary — a quick retry often fixes it. But when it lingers, it can indicate deeper communication issues between servers.

503: Service Unavailable — "Server's Taking a Nap"

A **503 Service Unavailable** usually means the server is overloaded, under maintenance, or just can't handle any more requests right now.

Think of it like calling customer support and hearing, "All our agents are currently busy. Please hold."

When big events happen — like concert ticket releases or Black Friday sales — websites often get hit by so many users that their servers buckle under the pressure.

Some smarter systems put up a waiting page or use queueing services. Others just faint and show a 503.

504: Gateway Timeout — "Nobody Answered the Phone"

This one's close cousins with the 502, but the message is slightly different.
A **504 Gateway Timeout** happens when one server tried to reach another, but the second one didn't respond in time.

Maybe it's down. Maybe it's just too slow.
It's like you calling your friend, waiting 30 seconds, and hanging up after too many rings.

400: Bad Request — "Speak Properly, Please"

A **400 Bad Request** means your browser sent something the server didn't understand.
It could be a corrupted cookie, malformed URL, or broken syntax in your request.

In simpler terms: you mumbled, and the server said, "Huh? Try again."

Clearing your cookies or cache often fixes this — it's the Internet's version of "turn it off and on again."

401: Unauthorized — "Who Are You?"

A **401 Unauthorized** error pops up when a site needs authentication — like a username and password — and you didn't provide it (or you did it wrong).

Picture knocking on a club door, and the bouncer saying, "Name's not on the list."

Once you log in properly, the gate usually opens. Unless you're still using the password "password123," in which case… please don't.

DNS Errors — "Can't Find That House"

When your browser says **"Server not found"** or **"DNS_PROBE_FINISHED_NXDOMAIN,"** it's not lying — it literally couldn't find the domain's IP address.

Either:

- The website no longer exists.
- The DNS records haven't propagated.
- Your Wi-Fi router is being dramatic again.

It's like asking for directions to a city that was deleted from the map.

SSL Errors — "Your Connection Isn't Private"

Modern sites use **HTTPS**, which encrypts communication between your browser and the server. But if the site's certificate is expired or misconfigured, your browser waves a big red flag.

Chrome loves to warn you with:

> "Your connection is not private. Attackers might be trying to steal your information."

Translation:

> "This website forgot to renew its safety badge. Proceed with caution (or not at all)."

It's not always malicious — sometimes a legitimate business just let its SSL certificate at Truehost.com expire. You can email them and tell them to stop being stingy, renew their SSL at Truehost. But that tiny lock icon exists for a reason: it means your data's safe.

Timeouts and Infinite Loading — "The Internet Is Tired"

Sometimes you don't get an error message at all — just a spinning wheel of doom.
That's a **timeout** — your browser sent the request, but the response took too long.

It's like texting someone who reads your message and never replies.
The Internet equivalent of being *ghosted.*

Why These Errors Matter

Here's the secret: every error you see is actually a sign the Internet *is working.*
It means your request traveled, reached somewhere, and got a response — even if that response was "nope."

If the Internet *didn't* work, you'd just see nothing at all. So, when you get a 404 or a 500, smile. Somewhere out there, servers are talking, routers are routing, and systems are doing their best to interpret your digital wish — they just tripped on a cable along the way.

2.8 Anatomy of a Web Request

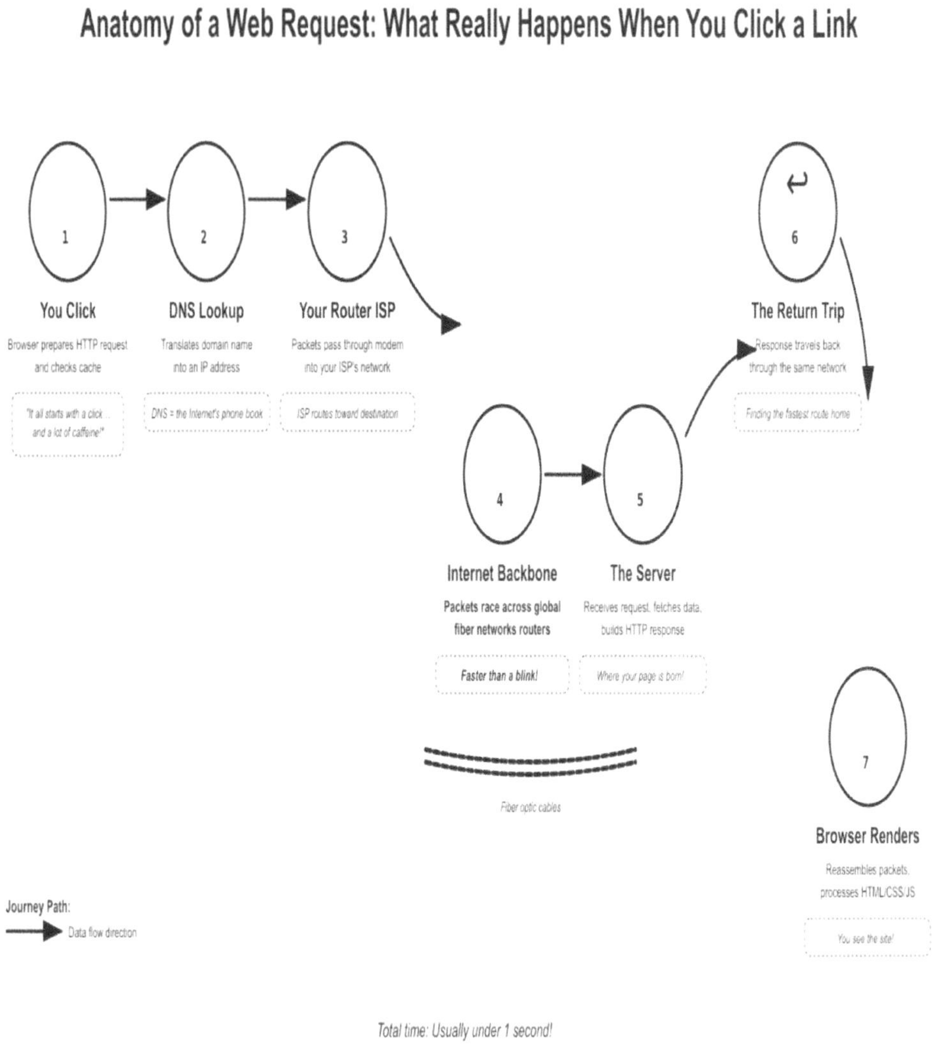

From click to screen in under one second — a global relay race of routers, servers, and code that makes the Internet feel like magic.

2.9 Summary & Bridge to Chapter 3 – From Click to Cloud

Congratulations — you've just followed one of the most complicated processes in human history… without touching a single line of code. 🎉

Every time you click a link, an invisible ballet unfolds. Your browser, your computer, your router, your ISP, DNS servers, global backbones, and remote data centers all work together in perfect (or sometimes chaotic) harmony — just to deliver a page full of memes, news, or cat videos in under a second.

The fact that it *feels* instant is the greatest magic trick technology ever pulled.

Let's Recap the Journey

1. The Click:
It all starts with your browser — the digital butler waiting for instructions. One click tells it to go fetch something from the web.

2. The Request:
Your browser breaks down your command into packets, wraps them in HTTP headers, and sends them into the network with directions to their destination.

3. DNS Lookup:
Since computers speak in numbers, DNS acts as the translator, turning human-friendly names like truehost.com into IP addresses.

4. Through the ISP:
Your request leaves home through your router and modem, enters your ISP's local network, and then hops across the Internet's global backbone — bouncing between routers faster than you can blink.

5. The Server:
Eventually, the packets reach a web server that reads your request, retrieves or generates the page, and sends it all back as an HTTP response.

6. Rendering the Page:
Your browser then takes the raw data (HTML, CSS, JavaScript) and paints it onto your screen — a process so fast and intricate it makes Formula 1 pit crews look lazy.

---

7. The Whole Cycle Repeats — Constantly.
Every link, every image, every scrolling feed is another micro version of this same process. You're clicking hundreds of times a day — and each one quietly triggers a world-spanning journey.

Why It Matters

Understanding what happens when you click a link turns the Internet from a black box into a knowable system.
You stop thinking of "the cloud" as something mystical floating above your head, and start recognizing it as **real machines, real people, and real processes** working in concert.

That awareness gives you:

- A better sense of *how things actually work*,
- The vocabulary to talk intelligently about technology,
- And maybe, a little awe for how fragile — and yet resilient — this global system truly is.

Because make no mistake: the Internet isn't just wires and code. It's one of humanity's largest cooperative projects — billions of devices speaking the same digital language, without any single boss in charge.

Summary Table

Stage	What Happens	Example
1. Click	You request a webpage.	Clicking a news article.
2. DNS	Domain translated to IP.	cnn.com → 151.101.1.67
3. ISP	Data leaves your local network.	Router → Modem → Fiber.
4. Internet Backbone	Packets hop across routers globally.	Dallas → Chicago → New York → London.
5. Server	Web server fetches data and responds.	Apache, Nginx, or AWS Lambda.
6. Browser Rendering	Your browser builds and paints the page.	HTML → CSS → JavaScript → Screen.

Quickfire Review: The Internet's Click-to-Show" Pipeline

So What About "The Cloud"?

You've probably noticed we keep hinting at "the cloud" — the mysterious place where all your photos, emails, and documents supposedly live.
Well, spoiler: *the cloud isn't a place.* It's just someone else's computer.

But not just one. Thousands — packed inside enormous data centers scattered across the planet.

Those data centers are the Internet's physical body. They host the servers that respond to your clicks, store your backups, power your favorite apps, and keep websites online 24/7.

If the Internet is a brain, data centers are the neurons — humming with activity, passing signals nonstop.
And in Chapter 3, you'll meet them properly.

Coming Up Next – Chapter 3: Understanding Servers & DNS

In Chapter 3, we'll take everything you've learned and zoom in on the *infrastructure* side of the story.
You'll discover:

- What a server really looks like (spoiler: not fluffy).
- How hosting works — shared, VPS, and cloud.
- Why "the cloud" is both marketing genius and engineering marvel.
- How domains, IPs, and DNS records tie it all together.
- What happens inside data centers, and why they're the Internet's beating heart.

By the end, you'll not only understand *how* the Internet moves — you'll know *where* it lives.

So take a sip of water, stretch your fingers, and get ready to open the door to the cloud itself. ☁□

Because if Chapter 2 showed you **the journey**, Chapter 3 will show you **the home.**

✅ **End of Chapter 2 — "What Happens When You Click a Link."**

CHAPTER 3 – UNDERSTANDING SERVERS & DNS

Sections:

3.1 What Exactly *Is* a Server?
3.2 From Your Laptop to the Cloud: Types of Hosting
3.3 What "The Cloud" Really Means
3.4 Inside a Data Center
3.5 Domains, IP Addresses & DNS in Action
3.6 How Websites Stay Online 24/7
3.7 Security, Backups & Redundancy
3.8 📊 *Diagram Placeholder – "The Cloud: Servers, Domains & You"*
3.9 Final Checklist – "Do You Really Understand the Internet Now?"
3.10 Glossary – "The Internet in Plain English"

3.1 What Exactly *Is* a Server?

The word "server" sounds grand — mysterious, maybe even intimidating — but at its core, a **server is just a computer** that serves.
That's literally why it's called that.

A server "serves" data, websites, videos, or files to anyone who requests them.
If your laptop is a personal chef cooking just for you, a server is a 24-hour restaurant feeding the entire city.

Same Ingredients, Different Kitchen

Under the hood, servers aren't very different from your everyday computer.
They have the same basic parts:

- **CPU:** The brain doing the calculations.
- **RAM:** The short-term memory for quick data access.
- **Storage:** Where files live (often in RAID arrays for safety).
- **Network Interface:** The connection to the Internet highway.

What makes a server special is how it's built and used:

- It runs nonstop — 24 hours a day, 365 days a year.
- It's optimized for stability, not selfies.
- It can handle hundreds or thousands of users at once.

No one logs in to watch movies or play games on it (unless that's literally its job — hello, Netflix). Its entire purpose is to respond to other machines' requests as fast as possible.

The "Headless" Workforce

Most servers don't even have monitors or keyboards. They live in racks, controlled remotely by administrators.
Picture a wall of blinking lights in a chilly room — that's the Internet's version of an office cubicle.

Technicians connect through **SSH (Secure Shell)** or remote dashboards to install updates, restart services, or check performance.

If something crashes, they don't drive to the data center to press "Ctrl + Alt + Del."
They reboot it remotely or let automation scripts heal the system.

Servers are the introverts of the tech world — happiest when left alone to hum quietly in the background.

One Machine, Many Roles

Depending on configuration, one server can wear multiple hats:

Role	What It Does	Real-World Example
Web Server	Hosts websites and handles HTTP requests.	Apache, Nginx, LiteSpeed.
Database Server	Stores structured information.	MySQL, PostgreSQL.
File Server	Manages shared files and backups.	Google Drive, Dropbox.
Mail Server	Sends and receives emails.	Gmail, Outlook servers.
DNS Server	Translates domain names to IP addresses.	Cloudflare DNS, Google Public DNS.

In big setups, each role gets its own machine (or cluster).
In smaller setups, one server may handle several jobs — a bit like a small-town diner where the cook also waits tables and washes dishes.

Physical vs. Virtual Servers

Here's where modern magic begins.

Originally, servers were **physical machines** — actual boxes stacked in racks. Then came **virtualization**, which lets one physical server split into many "virtual servers," each behaving like a separate computer.

It's like cutting a cake into slices — each slice is its own little server, isolated and independent, but all share the same base ingredients.

This made hosting cheaper, more flexible, and more scalable. Need another server? Just spin up a new virtual one instead of buying new hardware.

Virtualization paved the way for **cloud computing**, which we'll explore next.

A Peek Behind the Curtain

When you visit a website like truehost.com, your browser connects to one of Truehost's web servers. That server receives your request, fetches data (maybe from a database server), and sends the result back through the Internet — just as we saw in Chapter 2.

Multiply that process by millions of users and thousands of websites, and you start to see why servers are the unsung heroes of the Internet. They never get credit, but without them, "the cloud" would just be… vapor.

3.2 From Your Laptop to the Cloud: Types of Hosting

You could technically run a website from your laptop.
In fact, early web pioneers did exactly that — plug in a phone line, install a web server, and *boom*, you're live.

But then someone else tried to visit the site while the laptop was off, or streaming music, or overheating.
You guessed it, chaos!

That's why we have **hosting** — renting a slice of someone else's always-on, always-connected machine like Truehost to keep your site online even while you're asleep.

Hosting 101: A Digital Neighborhood

Think of the Internet like a city:

- **Websites** are houses.
- **Servers** are the plots of land.
- **Hosting providers** are the landlords.

When you "host" a website, you're renting a little digital real estate so visitors can reach you any time.

1⃣ Shared Hosting – The Apartment Complex

This is where most websites start — cheap, convenient, and surprisingly cozy.

Dozens or even hundreds of websites share the same physical server and resources (CPU, memory, storage). Each has its own "unit," but they all use the same electricity, plumbing, and hallway Wi-Fi.

Pros:

- Super affordable (a few dollars a month).
- Easy to set up — perfect for beginners.
- Maintenance handled by the hosting company.

Cons:

- Limited resources; your noisy neighbor can slow you down.
- Not ideal for high-traffic or complex sites.
- Less control — you can't install every software you want.

Best for: Small blogs, portfolios, or business landing pages.

If your site were an apartment, shared hosting is the studio with great rent but paper-thin walls.

2️⃣ VPS Hosting – The Condo Upgrade

VPS (Virtual Private Server) hosting is the next step up.
One physical machine is divided into multiple virtual servers using **virtualization** — each acts like an independent computer with its own operating system, CPU share, and memory allocation.

Pros:

- Dedicated resources, no noisy neighbors.
- More control — you can customize configurations.
- Scales better as traffic grows.

Cons:

- Costs more than shared hosting.
- Requires some technical know-how.

Best for: Growing businesses, e-commerce sites, or developers running apps.

VPS hosting feels like owning a condo: you share the building but control your own space. Decorate however you want — just don't knock down structural walls.

3️⃣Dedicated Servers – The Mansion

Now we're talking luxury.
A **dedicated server** means the entire machine belongs to you — all the CPU power, memory, and bandwidth are yours alone.

Pros:

- Maximum performance and customization.
- Highest level of security and reliability.
- Ideal for enterprise-scale apps or high-traffic sites.

Cons:

- Expensive (hundreds per month).
- You manage everything — updates, security, and troubleshooting.

Best for: Large companies, streaming platforms, or mission-critical systems.

It's the digital equivalent of owning your own estate — massive, private, and fully your responsibility when the roof leaks.

4️⃣ Cloud Hosting – The Smart City

Enter the modern era: **Cloud Hosting**.

Instead of one physical server, your site runs on a *network* of interconnected machines across multiple data centers. If one server fails, another instantly takes over.

Pros:

- Scalability — add resources instantly when traffic spikes.
- Redundancy — no single point of failure.
- Pay-for-what-you-use pricing.

Cons:

- Can get expensive at scale.
- Requires setup knowledge or managed services.

Best for: Modern web apps, SaaS platforms, global businesses.

Cloud hosting is like living in a futuristic smart city — your home floats across many interconnected neighborhoods, adjusting its size automatically when guests arrive.

5️⃣ Serverless Computing – Renting by the Second

For advanced users, the cloud offers an even newer model: **serverless computing**.
Despite the name, there *are* still servers — you just never have to manage them.

You upload your code to a platform (like AWS Lambda, Google Cloud Functions, or Azure Functions), and the provider automatically runs it whenever someone triggers it.

You pay only for the milliseconds your code actually runs.
No idle time. No maintenance. Just instant, scalable execution.

It's like using a taxi instead of owning a car — you pay only when you ride.

6⃣ Managed Hosting – The Concierge Service

Finally, there's **managed hosting**, where experts handle everything for you — updates, security, backups, performance tuning.

This can apply to any of the types above (shared, VPS, or cloud).
You just say, "I want my site online," and your host says, "Don't worry, we'll take care of it."

It's popular for businesses that want power without the headaches of server administration.

A Quick Comparison

Type	Analogy	Control	Cost	Ideal For
Shared	Apartment	Low	$	Beginners, blogs
VPS	Condo	Medium	$ $	Small businesses
Dedicated	Mansion	High	$ $ $	Large sites
Cloud	Smart City	Variable	$ – $ $ $	Scalable apps
Serverless	Taxi Service	Minimal	Pay-per-use	Developers
Managed	Hotel Concierge	Low (outsourced)	$ – $ $	Non-technical users

---

The Hosting Evolution

1. **1990s:** Everyone had a physical server — usually in a closet.
2. **2000s:** Shared and dedicated hosting became mainstream.
3. **2010s:** Virtualization and cloud computing took over.
4. **2020s:** Serverless and distributed architectures rule.

Today, even small startups can deploy systems once reserved for corporations — thanks to the cloud's flexibility.

Where You Fit In

When you buy a hosting plan from companies like Truehost, Bluehost, or AWS, you're really renting computing power.
Whether you get a small apartment (shared hosting) or a skyscraper suite (cloud instance), it all lives inside a data center somewhere — humming away 24/7, waiting for requests like yours.

And that brings us to the real heart of "the cloud" — where all these servers actually live.

3.3 What "The Cloud" Really Means

People love saying *"It's in the cloud,"* as if the Internet were a fluffy celestial storage unit floating above our heads.
Reality check: the cloud isn't in the sky.
It's in **massive warehouses full of servers** — heavy, humming, humming machines stacked on racks, cooled by industrial fans, and powered by more electricity than a small town.

When you upload a photo, stream a show, or send an email, your data doesn't go *up* — it goes *sideways,* into one of these global data centers.

So let's clear the fog once and for all.

The Cloud = Someone Else's Computer

At its simplest, *the cloud* is just **a network of computers (servers) that you rent instead of own.**

In the old days, if you wanted to run a website or an app, you had to buy a physical server, install software, and maintain it yourself.
If you needed more power, you had to buy more machines — and when traffic dropped, those extra servers sat there collecting dust.

Cloud computing flipped that model completely.
Instead of owning servers, you **lease computing power on demand** from providers like Amazon Web Services (AWS), Microsoft Azure, Cloudoon or Google Cloud Platform (GCP).

You pay only for what you use — CPU time, memory, storage, bandwidth — like utilities.
If your app goes viral overnight, you can scale instantly. If it flops (sorry), you scale down and stop paying. That flexibility is the superpower of the cloud.

The Three Layers of Cloud Computing

Cloud computing comes in three main flavors, or "service models." Think of them as layers of convenience — from building your own kitchen to ordering takeout.

Model	What You Get	Analogy
IaaS – Infrastructure as a Service	Virtual servers, storage, and networking. You manage the software.	Renting an empty apartment. You bring the furniture (apps).
PaaS – Platform as a Service	Preconfigured environments for developers. You just deploy code.	Renting a furnished apartment.
SaaS – Software as a Service	Finished apps you use directly (email, CRMs, etc.).	Staying at a hotel — everything's ready.

Examples:

- **IaaS:** AWS EC2, Google Compute Engine, Cloudpap, DigitalOcean.
- **PaaS:** Heroku, Firebase, Truehost, Vercel.
- **SaaS:** Gmail, Zoom, Dropbox, Canva, Olitt, Netflix.

Every cloud product you've ever used fits somewhere on that spectrum.

How the Cloud Scales Like Magic

Remember Chapter 2's description of packets bouncing between routers?
In the cloud, there are millions of those packets — constantly routed between virtual machines inside enormous server farms.

Each data center contains thousands of **physical servers**, but software called a **hypervisor** divides them into millions of **virtual machines (VMs)** or **containers**.
When demand spikes, new instances are spun up automatically within seconds.Let's say you run a web app that suddenly gets 50× more users overnight.

- On physical hardware, your site would crash.
- In the cloud, the system detects the load, clones your app across multiple servers, and distributes the traffic seamlessly.

When the surge passes, it automatically scales down again.

That's the beauty of **elastic computing** — resources that grow and shrink with your needs.

Regions, Zones, and Redundancy

The cloud isn't one giant building. It's a **global network** of data centers, divided into regions and availability zones.

- **Region:** A geographical area (e.g., "US-East (Virginia)" or "Europe-West (London)").
- **Zone:** A cluster of data centers within that region — each independent but connected.

This setup ensures **redundancy** — if one zone fails (say, due to a power outage), traffic automatically reroutes to another.
Your app doesn't even blink.

For example, Amazon Web Services runs over **100 availability zones** across **30+ regions** worldwide.
Google and Microsoft operate similar networks.
Together, these form the modern Internet's nervous system — the physical backbone behind the "cloud" buzzword

---

Cloud Storage: Your Data's New Home

When you save something "to the cloud" — photos, backups, documents — it's stored on cloud storage systems like:

- **Amazon S3 (Simple Storage Service)**
- **Google Cloud Storage**
- **Microsoft Azure Blob Storage**
- **Cloudoon S3**

These systems don't just store files in one place.
They automatically replicate your data across multiple drives, racks, and sometimes continents — ensuring durability even if hardware fails.

So if lightning strikes a data center in Oregon, your selfies are still safe thanks to a mirrored copy sitting peacefully in Virginia.

Edge Computing: The Cloud Moves Closer

As more users demand real-time speed (for gaming, video calls, and IoT), cloud providers are pushing the edge — literally.

Edge computing brings cloud power physically closer to you by placing smaller servers in regional data centers or even cell towers.
Instead of your data traveling 1,000 miles, it might only go 10.

This reduces latency — the time between click and response — and makes things like streaming, self-driving cars, and augmented reality possible.

It's like having a mini-cloud in your neighborhood instead of waiting for the big one across the ocean.

Why Businesses Love (and Fear) the Cloud

Pros:

- ❢ **Scalability:** Add resources instantly.
- 💰 **Cost efficiency:** Pay only for what you use.
- ⚙□ **Automation:** No hardware headaches.
- □ **Global reach:** Deploy apps near your users.
- 🔒 **Security:** Enterprise-grade protections most small firms couldn't afford alone.

Cons:

- □ **Costs can surprise you.** If you don't monitor usage, your bill can skyrocket faster than your traffic.
- 🔐 **Shared responsibility.** The provider secures the infrastructure, but you still have to secure your own apps.
- □□ **Outages happen.** Even the biggest clouds occasionally go down — remember, it's still just computers.

In short: the cloud gives you superpowers, but you still need to use them wisely.

Fun Fact: Cloud vs. Fog

There's even a trend called **"fog computing"** — basically, the cloud pushed closer to the ground (like edge computing for IoT devices).
Your smart fridge, thermostat, or security camera might process some data locally before syncing with the cloud.
So yes, even your fridge is technically a "fog node." Welcome to the future.

So Where *Is* the Cloud?

If you could visit one, you'd find row after row of metal racks filled with blinking servers, whirring fans, and miles of fiber cables.
Each machine runs quietly — responding to millions of clicks, uploads, and queries every second.

They live everywhere:

- Oregon, Iowa, and Virginia (U.S.)
- Frankfurt and Dublin (Europe)
- Singapore, Tokyo, and Sydney (Asia-Pacific)
- Nairobi, Lagos, Johannesburg (Africa, finally growing fast!)

Each one is a tiny piece of the global puzzle that makes the Internet "just work."

In One Sentence

> "The cloud is not somewhere *up there* — it's millions of computers *down here*, working together so you never have to think about them."

3.4 Inside a Data Center

If the Internet had a body, **data centers** would be its beating heart. Every photo you upload, every Netflix show you stream, every website you visit — it all lives in one of these giant, humming buildings somewhere on Earth.

You might drive past one and not even know it. From the outside, a data center looks like a minimalist warehouse — no windows, no signs, just heavy security and an endless low hum. But inside? It's the digital equivalent of a spaceship.

Let's open the (very secure) doors and take a tour.

1️⃣ The Physical Layout – Racks, Rows, and Redundancy

The moment you step in, the first thing you'll notice is **rows upon rows of metal racks** filled with servers — each rack containing 40 to 60 machines stacked like pizza boxes.

Each server is plugged into:

- Power circuits with redundant backup lines
- High-speed fiber network connections
- Cooling ducts or cold air aisles

The floor layout follows a **"hot aisle / cold aisle"** design:

- **Cold aisles:** Rows where chilled air is pumped in to cool the servers' intake vents.
- **Hot aisles:** Rows where warm air is expelled and whisked away by ventilation systems.

This pattern prevents hotspots and keeps temperatures stable. It's an elegant choreography of airflow.

2️⃣ Power: The Lifeblood of the Internet

Data centers consume *massive* amounts of electricity.
Some facilities use as much power as 50,000 homes.

Every server, router, and cooling fan needs a continuous energy supply, and even a one-second blackout could mean thousands of websites going offline.

That's why most data centers have:

- **Dual power feeds** from separate utility grids
- **Battery backup systems (UPS)** for instant protection
- **Diesel generators** that can run for days if the grid fails

When you see headlines about tech giants investing in solar farms or wind energy — this is why.
Keeping the Internet running takes *a lot* of juice.

3️⃣Cooling: Fighting the Heat Monster

Servers generate serious heat. Without proper cooling, they'd literally melt. That's why data centers spend almost as much energy on cooling as they do on computing.

Common cooling methods:

- **CRAC (Computer Room Air Conditioning)** units that circulate chilled air.
- **Liquid cooling** — pipes running cool water near CPUs.
- **Outside air economization** — bringing in cold outdoor air in cooler climates.
- **Hot/cold aisle containment** — keeping warm and cool air separate to maximize efficiency.

Modern data centers measure efficiency using **PUE (Power Usage Effectiveness)** — the ratio of total facility energy to computing energy. A perfect score is 1.0 (meaning all energy goes directly to computing). Most high-end facilities achieve around **1.1–1.3**, which is incredibly efficient.

Some even use *underwater data centers* or build in cold climates like Finland or Wyoming to take advantage of natural cooling. (Yes, Microsoft really sank servers into the ocean once — and it worked.)

4️⃣Connectivity: The Internet's Superhighway

Every data center is connected to the rest of the world via **fiber-optic cables** — thin glass strands that transmit light pulses at nearly the speed of light.

These cables form the Internet's **backbone**, linking major data hubs across continents and under oceans.

A single data center might have connections to multiple **Internet Exchange Points (IXPs)**, ensuring redundancy.
If one cable is cut (by accident or, yes, by a shark — it's happened), traffic automatically reroutes through another.

Routers and switches inside the facility handle this data routing. They ensure every packet finds the fastest, safest path to its destination — just like air traffic controllers managing digital planes.

5️⃣ Security: Where the Real Fort Knox Lives

If you think airport security is tight, you should see a hyperscale data center. Getting inside one involves multiple layers of protection:

- **Biometric scanners** (fingerprint, facial, or even retina scans)
- **Security guards** on site 24/7
- **Mantraps** (two-door systems that only open one at a time)
- **Video surveillance** in every corridor
- **Visitor logs and keycard access levels**

Many facilities also enforce **"no camera, no phone"** rules.
Technicians use secure, monitored devices for maintenance.
After all, this is where governments, banks, and Fortune 500 companies keep their data — one breach could mean billions in losses.

6□Fire Suppression: The Quiet Guardians

Fires are a nightmare for any data center — but water and electronics don't mix.
So instead of sprinklers, most use **gas-based suppression systems** like FM-200 or Novec 1230, which smother flames without harming hardware.

Sensors detect smoke or heat changes instantly, and the gas deploys in seconds.
It's one of the few places on Earth where you might pray *not* to see water if something's burning.

7 The Staff: The Human Element

Behind all the automation, data centers still need people — though surprisingly few.
A massive 100,000-square-foot facility might have only a couple dozen engineers on-site, supported by remote monitoring teams.

These technicians check:

- Rack health and power usage
- Cooling systems and airflow
- Network traffic patterns
- Security and environmental sensors

They're like digital farmers tending vast fields of machines. Everything's automated, but they're always ready to fix what the bots can't.

8️⃣ Sustainability: Greening the Cloud

The big cloud providers — Google, Amazon, Microsoft, Meta — are racing to make data centers more sustainable.
Initiatives include:

- **Renewable power** (wind, solar, hydroelectric).
- **Heat recycling** — using waste heat to warm nearby homes.
- **AI-based cooling systems** that adjust in real time.
- **Carbon offsets** and "net-zero" commitments.

Google claims many of its data centers already run on **100% renewable energy**, and others are aiming for **24/7 carbon-free operations** this decade.
The cloud, once energy-hungry, is slowly learning to diet.

9️⃣ Hyperscale and Edge Facilities

Not all data centers are equal.

- **Hyperscale** facilities (run by AWS, Google, Azure) can hold hundreds of thousands of servers.
- **Colocation centers** rent out smaller spaces to businesses (like Truehost or Digital Realty).
- **Edge data centers** are compact versions deployed near cities to reduce latency.

Whether massive or miniature, they all serve the same purpose: keep the Internet alive, stable, and always available.

A Final Thought

When you send a WhatsApp message or upload a YouTube video, remember: that data ends up somewhere — *a real, physical place.*

Rows of humming servers blink quietly, thousands of miles away, faithfully holding your memes, your emails, your company records, and your late-night online shopping regrets.

That's the "cloud." Not magical. Just magnificent.

3.5 Domains, IP Addresses & DNS in Action

You now know what servers are, how hosting works, and where "the cloud" actually lives.
But there's one more puzzle piece that makes the Internet usable: **how we find things.**

Because even if your website sits in the most high-tech data center in the world, nobody's visiting it unless they know where to look.

That's where **domains, IP addresses, and DNS** team up to make everything work seamlessly — like a digital GPS for the entire Internet.

1️⃣IP Addresses: The Internet's Street Numbers

Every device connected to the Internet — your phone, laptop, or server — has a unique **IP address** (short for Internet Protocol).
Think of it as a postal address for your machine.

Example:

- IPv4 address: 93.184.216.34
- IPv6 address: 2606:2800:220:1:248:1893:25c8:1946

Without these numbers, computers wouldn't know where to send your requests.

When your browser fetches a webpage, it uses the IP to route packets between devices — just like the postal service uses ZIP codes to deliver mail.

IPv4 vs IPv6: Running Out of Numbers

The original system, **IPv4**, allows for about 4.3 billion unique addresses. That sounded infinite in the 1980s — until every smartphone, smartwatch, smart toaster, and smart cat feeder needed one.

We simply ran out.

That's why the Internet adopted **IPv6**, which supports a mind-boggling **340 undecillion** addresses (that's 340 followed by 36 zeros). Enough for every grain of sand on Earth to have its own IP — and a backup.

Today, most major providers (Google, Cloudflare, ISPs) use both systems in parallel. Your computer quietly handles the translation so you don't have to memorize anything longer than your credit card number.

2□Domain Names: The Human Shortcut

Let's face it — no one wants to type "93.184.216.34" to visit a website. That's why we have **domain names** — human-friendly labels for those long, unfriendly IPs.

When you type www.truehost.com, your browser doesn't actually understand the name. It immediately calls on DNS (the Domain Name System) to translate it into an IP address — a process called **name resolution**.

This is where the Internet's secret phonebook gets to work.

3□How DNS Works in Real Life

We covered the basics of DNS back in Chapter 2, but now let's connect it to your shiny new understanding of hosting and servers.

Here's what happens step-by-step when someone visits your site, olitt.com:

Step 1: User Request

A user types olitt.com into their browser.
The browser asks the operating system:

> "Do we already know this address?"

If it's cached, the answer comes instantly. If not, it triggers a full DNS lookup.

Step 2: Recursive Resolver

The request goes to a **recursive resolver** — typically run by your ISP or a public DNS like Google's 8.8.8.8.
Its job? To find the answer, no matter how many phone calls it has to make.

Step 3: The Root Name Server

If the resolver doesn't know the IP, it contacts a **root server**.
There are 13 root server clusters worldwide (each replicated hundreds of times).
The root says, "I don't know that exact site, but it's a .com domain — ask the .com directory."

Step 4: The TLD Name Server

Next stop: the **Top-Level Domain (TLD)** name server for .com.
This server replies:

> "Ask the authoritative name server for olitt.com — it knows the real answer."

Step 5: The Authoritative Name Server

This is the final authority — the DNS server controlled by your hosting provider or domain registrar.
It looks up your domain's **A record** (the mapping from domain → IP) and says:

> "Here you go: 203.0.113.17"

Step 6: Resolution Complete

The resolver passes the answer back to your browser, which now knows exactly where to send the HTTP request.

All of that happens in about 100 milliseconds — faster than you can blink.

4️⃣ The DNS Records That Make It All Work

A domain's DNS configuration includes multiple record types — each serving a different purpose:

Record	Function	Example
A	Maps a domain to an IPv4 address.	olitt.com → 203.0.113.17
AAAA	Maps to an IPv6 address.	Olitt.com → 2606:abcd::1
CNAME	Alias — points one domain to another.	www.olitt.com → olitt.com
MX	Mail Exchange — routes email.	Gmail uses this for @olitt.com inboxes.
TXT	Text data — used for verification, SPF, DKIM, etc.	"v=spf1 include:_spf.google.com ~all"
NS	Name Server — specifies which DNS servers are authoritative for your domain.	ns1.cloudoon.com

Together, these tiny records act like the Internet's Rolodex — telling everyone who's who and where to find them.

5️⃣How Domains Tie to Hosting

When you register a domain (say, with Truehost), it doesn't automatically "live" anywhere yet.
You must point its DNS records to your web host's IP — effectively telling the Internet,

> "Hey, this domain's website is served from that server over there."

Example:

- Domain: olitt.com (registered at Truehost)
- Hosting server IP: 104.21.12.145
- DNS A record: olitt.com → 104.21.12.145

Now, whenever someone types your domain, their browser knows exactly which server to visit.

Change your web host? Just update the DNS records.
Your domain stays the same, but the "address" moves — like forwarding your mail to a new home.

6□Propagation: The Internet's Gossip Network

When you update DNS, the changes don't take effect instantly.
They have to **propagate** across DNS caches worldwide — from your local router to your ISP to global resolvers.

Depending on TTL (Time To Live) settings, that can take anywhere from a few minutes to 48 hours.
So if you've ever seen your new site appear for some people but not others — that's the Internet still "catching up."

7□Real-World Example: Visiting Netflix

Let's visualize it with something you actually use daily.

When you type www.netflix.com:

1. Your DNS resolver finds the closest edge server (Netflix runs thousands worldwide).
2. DNS points your request to the nearest node using **GeoDNS** — ensuring the fastest connection based on location.
3. Your browser connects to that edge node (not California or Ireland — probably a city near you).
4. The edge node delivers cached data or streams directly.

That's why Netflix loads instantly — it's already waiting close to you.

8️⃣Why This Matters

Understanding domains and DNS isn't just trivia — it's digital literacy.
It's knowing:

- How your emails reach their destinations,
- Why a site won't load when DNS breaks,
- How hackers sometimes hijack domains by altering DNS records.

When you grasp DNS, the Internet stops being magic and starts being **knowable**.

You realize that behind every .com lies a chain of machines talking politely, all agreeing where "home" really is.

In Short

Domains are names, IPs are numbers, and DNS is the translator that keeps them in sync — the Internet's invisible switchboard operator.

3.6 How Websites Stay Online 24/7

If you've ever wondered why your favorite website never seems to "sleep," here's the truth: it doesn't.
The Internet is always awake — somewhere in the world, someone's clicking, streaming, or doomscrolling.

Keeping websites online 24/7 isn't luck. It's an elaborate combination of **redundancy, automation, and backup systems** — all working together so you rarely see the chaos underneath.

Let's peel back the curtain and see what makes the web so… unstoppable.

1️⃣Uptime: The Internet's Pulse Rate

Every web host boasts about their **uptime**, usually written as a percentage —

> "99.9% uptime guaranteed!"

That sounds impressive, but what does it mean?

Uptime	Downtime Per Year
99%	~3.65 days
99.9%	~8.76 hours
99.99%	~52 minutes
99.999% ("five nines")	~5 minutes

Reaching 99.999% uptime is the Holy Grail — the digital equivalent of running a marathon without tripping once.

Behind those numbers are layers of systems ensuring that even when something fails, *something else immediately steps in.*

2️⃣ Redundancy: The Art of Never Depending on One Thing

The golden rule of uptime is simple:

> "Don't have a single point of failure."

Every major website uses **redundancy** — duplicate hardware, network paths, and data copies that kick in automatically when the main system fails.

There are multiple types:

- **Hardware redundancy:** Two (or more) power supplies, network cards, and drives in every server.
- **Network redundancy:** Multiple ISPs and routers, so traffic reroutes if one connection dies.
- **Power redundancy:** Battery UPS + diesel generators + dual electrical feeds.
- **Geographic redundancy:** Copies of your site running in multiple data centers around the world.

If one fails, another takes over instantly.
Your browser doesn't even notice — it just keeps loading the page.

3️⃣ Load Balancing: The Digital Traffic Cop

When a site has thousands (or millions) of visitors, one server can't handle it all.
Enter **load balancers** — specialized systems that distribute traffic evenly across multiple servers.

Picture a toll booth with several lanes.
The load balancer decides which lane (server) each car (user request) should take, based on:

- Current traffic load
- Server health
- Geographic proximity
- Response time

If one server slows down or crashes, the balancer automatically diverts new traffic elsewhere.
Users keep browsing without interruption.

Large-scale systems — like YouTube, Facebook, or Truehost Cloud — may use dozens of load balancers layered together, orchestrating billions of requests per day like a symphony of routers.

4️⃣Content Delivery Networks (CDNs): Speed via Local Copies

Earlier, we met **CDNs** — networks of servers strategically placed around the globe to store cached versions of popular content (images, videos, scripts).

If your main website lives in New York but someone in Nairobi visits, a CDN node in Kenya delivers the page instead of routing all the way to the U.S. That's not just faster — it also lightens the load on your origin server.

Popular CDNs include:

- **Cloudflare**
- **Akamai**
- **Amazon CloudFront**
- **Fastly**

CDNs are the reason global websites load in under two seconds, regardless of where you are.
They're also part of uptime — if your origin goes offline briefly, the CDN can often keep serving cached copies to users until it's back.

5️⃣ Monitoring: The Internet's Night Watch

Even with redundancy, things *still* break.
That's why monitoring systems exist — the Internet's tireless sentinels.

These tools continuously check:

- Is the website responding?
- Are pages loading correctly?
- Is the server temperature normal?
- Are there traffic spikes or strange activity?

If anything looks off, they alert engineers instantly via email, SMS, or even Slack.

Tools like **Pingdom**, **UptimeRobot**, **Datadog**, and **Zabbix** are common watchdogs.
Some even perform **synthetic testing**, mimicking user behavior to catch issues before real users notice.

The result? Problems are detected — and fixed — often before anyone even realizes something went wrong.

6️⃣ Backups: Plan B, C, and D

Let's say worst-case scenario: the whole system crashes.
Maybe a power failure, a flood, or someone accidentally deletes a production database (it's happened to billion-dollar companies).

That's why every serious operation keeps **backups** — and lots of them.

There are several kinds:

- **Full backups:** Everything saved in one copy.
- **Incremental backups:** Only new or changed files.
- **Off-site backups:** Copies stored in another data center or region.
- **Versioned backups:** Older versions preserved for rollback.

The "3-2-1 rule" is gospel:

> Keep 3 copies of data, on 2 types of media, with 1 stored off-site.

Backups ensure business continuity — so even when the unthinkable happens, websites rise from the ashes like digital phoenixes.

7️⃣ Auto Healing & Failover Systems

Modern cloud platforms like AWS and Google Cloud take redundancy to sci-fi levels.
If a virtual machine crashes, an **auto-healing system** automatically spins up a new one — sometimes within seconds.

Similarly, **failover clusters** detect outages and shift operations to healthy systems, often in another region.
For mission-critical systems (banks, hospitals, flight tracking), this isn't optional — it's survival.

In the background, scripts and orchestration tools (like Kubernetes) keep moving workloads around like digital chess pieces, always keeping enough pieces on the board to win.

8️⃣ Security & DDoS Protection

Another threat to uptime isn't just failure — it's *attack*.
Hackers use **DDoS (Distributed Denial of Service)** attacks to flood servers with junk traffic, overwhelming them until real users can't connect.

To defend against this, sites use:

- **Firewalls** that block malicious IPs.
- **Rate limiting** that throttles abusive traffic.
- **CDNs and reverse proxies** that absorb and filter attacks.

Cloudflare alone blocks **over 100 billion malicious requests per day** across its network — silently protecting millions of websites that would otherwise crash.

So yes, your cat video site might have been saved by a global cyber-shield at some point.

9️⃣ Humans + Automation: The Final Line of Defense

Despite all the AI, scripts, and backups, uptime still depends on human engineers.
They plan, test, and tweak systems constantly — patching vulnerabilities, improving response times, and preparing for anything.

But unlike the early Internet (where humans fixed things manually), most modern uptime work happens automatically.
AI systems predict hardware failure before it happens, move workloads preemptively, and even order replacement parts before anyone wakes up.

The machines are watching the machines.
It's equal parts terrifying and impressive.

In Short

Keeping a website online 24/7 is like running a concert where the lights can *never* go out.
It takes redundancy, automation, vigilance — and a lot of caffeine. ☕

So the next time a page loads instantly, take a second to appreciate the invisible army of machines and humans making it happen.

3.7 Security, Backups & Redundancy

You've seen how websites stay online 24/7.
But staying *online* is only half the battle — the other half is staying **secure** and **recoverable**.

Because when you have billions of people, zettabytes of data, and thousands of hackers with free time, things are bound to get messy.

That's why modern Internet infrastructure is built on three pillars:

1. **Security** — keeping bad actors out.
2. **Backups** — keeping your data safe.
3. **Redundancy** — keeping everything running even when stuff breaks.

Together, they form the Internet's version of insurance: invisible, expensive, but absolutely essential.

1️⃣ Security: The Internet's Armor

Every day, servers are bombarded with login attempts, spam bots, malware, and denial-of-service attacks.
Without strong defenses, even the smallest site could crumble within hours.

Let's look at the major layers of protection.

a. Firewalls — The Digital Gatekeepers

A **firewall** is like the bouncer at the front door of a club.
It decides who gets in and who doesn't, based on a set of rules.

Types of firewalls:

- **Network firewalls:** Filter traffic at the router level (blocks suspicious IPs).
- **Application firewalls (WAF):** Protect web apps by blocking malicious requests (like SQL injections).
- **Cloud-based firewalls:** Operate at scale — filtering traffic across entire networks (e.g., Cloudflare, AWS WAF).

Example: When a hacker's bot tries to brute-force your login, the WAF quietly drops the connection before your server even notices.

b. Encryption — Turning Data into Gibberish

Ever noticed the little padlock 🔒 next to a website's URL?
That means it's using **HTTPS**, secured by **SSL/TLS encryption**.

Encryption scrambles data into unreadable code as it travels between your browser and the server.
Even if someone intercepts it, all they see is nonsense.

There are two main types:

- **In-transit encryption:** Protects data as it moves (e.g., between your laptop and Gmail).
- **At-rest encryption:** Protects stored data on disks or cloud storage (e.g., AWS S3 encryption).

Together, they ensure your passwords, messages, and cat memes stay private.

c. Authentication & Access Control

Who can access what? That's where authentication comes in.

- **Passwords:** Still the weakest link (too many "123456" users).
- **MFA (Multi-Factor Authentication):** Adds another layer — like a code texted to your phone.
- **SSH keys & access roles:** Used in data centers to control who can manage which servers.

Even employees at big tech companies have limited access.
For instance, a Google engineer can't just peek into your Gmail inbox — systems are segmented by strict permissions and audit logs.

d. Intrusion Detection & Threat Monitoring

Even with locks on every door, you still need an alarm system.

Modern infrastructures use **IDS (Intrusion Detection Systems)** and **SIEM platforms (Security Information and Event Management)** to monitor all traffic in real time.

They flag:

- Strange login patterns
- Data exfiltration attempts
- Sudden spikes in CPU/network usage
- Unauthorized file changes

It's like having digital security cameras that never blink.

---

e. Patching & Updates

Most hacks exploit known vulnerabilities — bugs that already have fixes.
So the simplest security habit is also the most ignored: **update your software.**

From routers to browsers, every update often includes "security patches" that close those open doors.
Automation tools like **Ansible**, **Chef**, or **Odoo scripts** can deploy updates across hundreds of servers instantly.

When IT teams delay patches, it's like leaving your front door unlocked in a bad neighborhood.

2️⃣ Backups: The Internet's Safety Net

No matter how good your security is, accidents happen — hard drives fail, data gets corrupted, humans delete the wrong folder (yes, even at Google once).

That's why backups exist. They don't prevent disasters; they undo them.

Let's unpack how serious operators handle backups.

---

a. The 3–2–1 Backup Rule

The golden standard of data safety:

> **3 copies** of your data,
> **2 different media types**,
> **1 stored off-site.**

Example:

- One on your primary server.
- One on an external drive or NAS.
- One in a remote cloud region or separate provider.

This ensures that even if your data center floods, burns, or gets abducted by aliens, at least one copy survives.

---

b. Automated & Incremental Backups

Manual backups are fine for hobbyists.
Real systems use **automated schedules** that run daily (or hourly).

They're also **incremental**, meaning only new or changed files are copied — saving time and storage space.

Tools like **rsync**, **BorgBackup**, and **Veeam**, or cloud-native options like **AWS Backup** and **Google Snapshot**, handle this seamlessly.

---

c. Versioning & Snapshots

Ever wish you could rewind time? In IT, you can.

Backups often include **versioning** — letting you restore older versions of files before they were corrupted or overwritten.

In cloud systems, **snapshots** capture the exact state of a virtual machine or database at a moment in time.
If something breaks, you just roll back to a previous snapshot like pressing "Undo" on the Internet.

---

d. Off-Site & Cross-Region Replication

Backups stored in the same place as your main data are basically sitting ducks. That's why serious operators keep copies **off-site** — often in a completely different region or even country.

Cloud providers replicate your data automatically across multiple **availability zones**, sometimes thousands of miles apart.

Example: A backup stored in AWS "us-east-1" might be mirrored in "us-west-2."
Even if one coast goes down, the other coast keeps your website alive.

3️⃣Redundancy: Because Failure *Will* Happen

We touched on redundancy earlier, but it's so vital it deserves its own spotlight here.

Redundancy means designing systems that assume something *will* break — and planning accordingly.

Types include:

- **N+1 redundancy:** One extra component for every set of N. If one fails, the spare takes over.
- **Active-active clusters:** Two or more servers handle live traffic simultaneously — if one dies, the others keep running without delay.
- **Database replication:** Copies of the same database stay synced across regions.

Redundancy isn't glamorous — users never notice it when it works.
But when it fails, *everyone* notices.

4️⃣Real-World Example: Gmail's Invisible Fortress

Google's Gmail service runs across thousands of servers in dozens of data centers.
Every email you send is stored redundantly — across multiple machines, multiple disks, and multiple cities.

If one drive dies, nothing happens.
If an entire data center goes offline, another instantly picks up the load.
That's why you can't "crash Gmail" by accident — it's too distributed to fail in one place.

The same is true for AWS, Azure, and even smaller cloud hosts like Truehost — redundancy isn't optional; it's the business model.

5️⃣ Human Redundancy: The Overlooked One

Believe it or not, even IT teams plan for **people failure** — vacation, illness, burnout, or just someone forgetting their laptop password.

That's why they use:

- Shared credentials stored securely in vaults like **1Password** or **Bitwarden**.
- **Runbooks** (step-by-step recovery procedures).
- **Disaster recovery drills** (yes, they simulate "data apocalypse" days).

Because the Internet is run by humans — and humans occasionally spill coffee on their keyboards.

In Short

Security stops bad things.
Backups recover from bad things.
Redundancy ensures bad things don't ruin everything else.

These three together make the Internet what it is: **resilient, recoverable, and remarkably reliable** — even in a world full of chaos.

3.8 ▥The Cloud: Servers, Domains & You

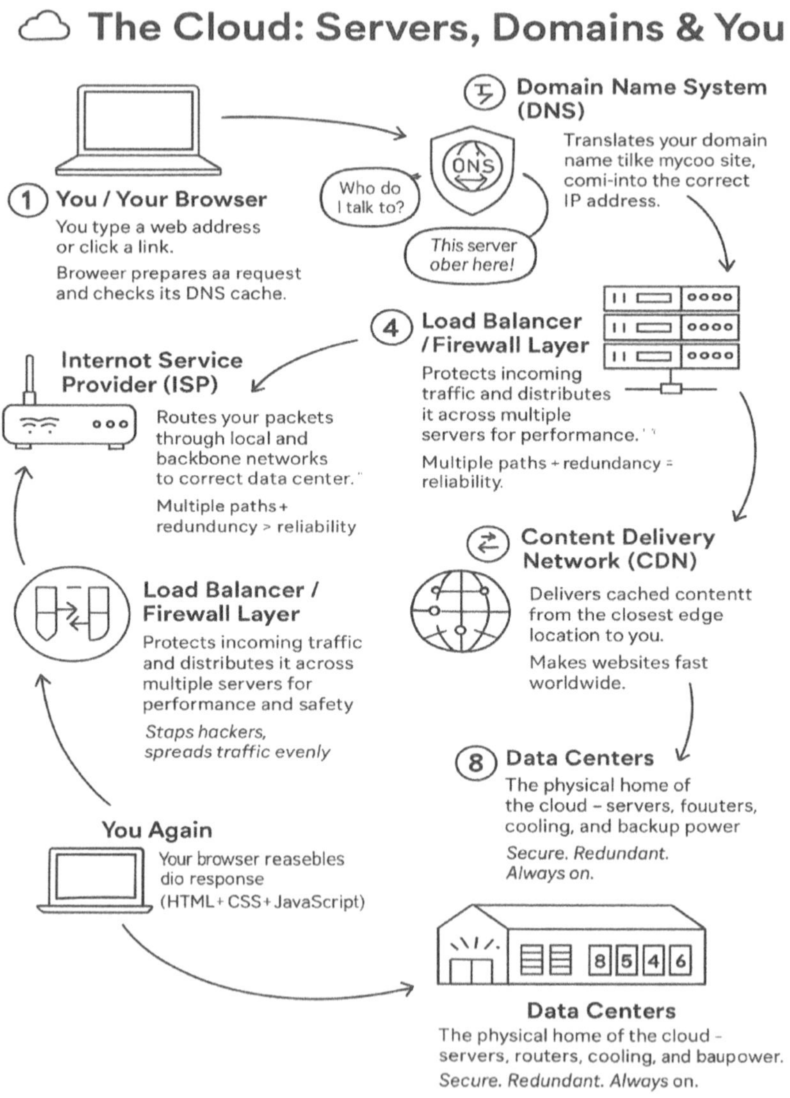

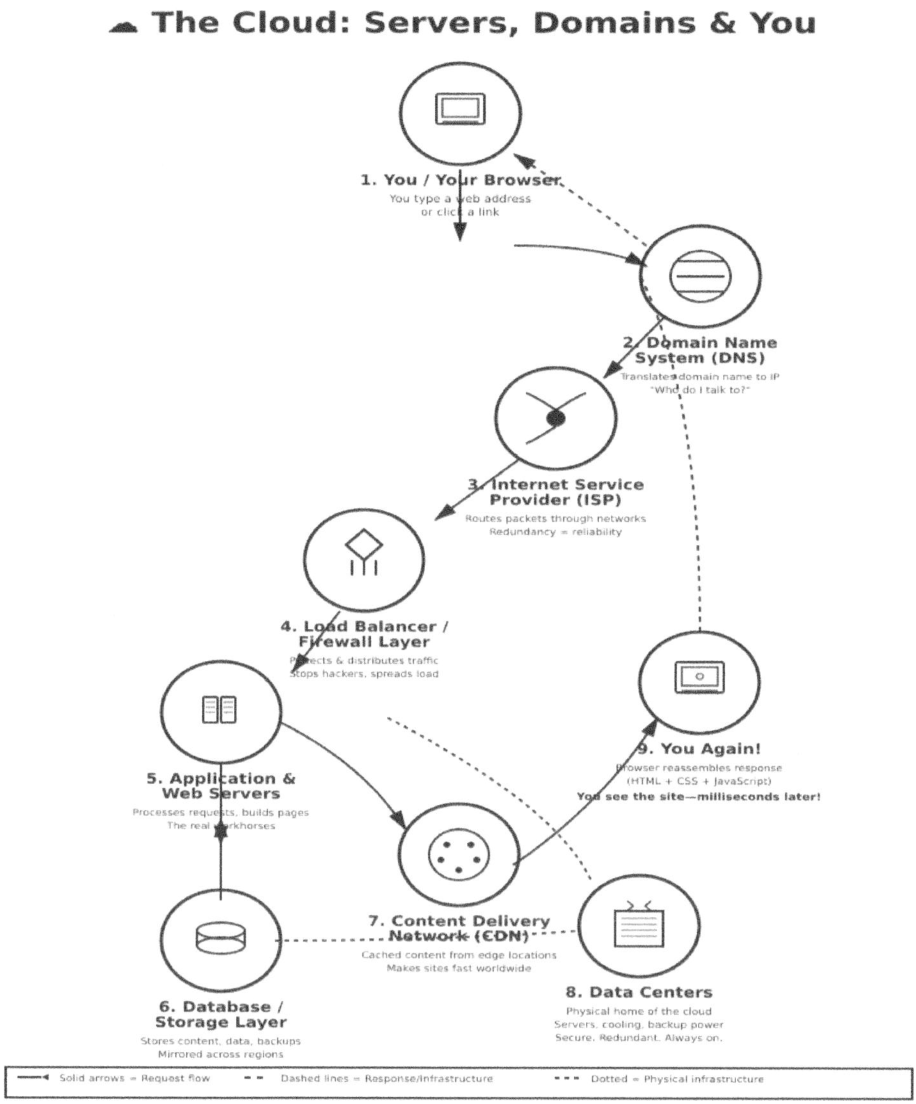

From click to cloud and back again — your data travels across continents, through cables, servers, and security layers, all working invisibly so the Internet feels instant.

3.9 ✅ Final Checklist – "Do You Really Understand the Internet Now?"

You've made it to the end!
Before we wrap this book and release you into the wild web with newfound wisdom, let's test how much of it *actually stuck*.

Grab a coffee, crack your knuckles, and go through this quick checklist.
If you can confidently check off most of these boxes, congratulations — you officially understand what "the cloud" means (and you're now slightly less likely to fall for tech jargon at parties).

☐ Section 1 – The Big Picture

☐ I can explain that the Internet isn't a single thing — it's a network of networks working together.

☐ I understand that the "cloud" is just a bunch of connected servers in data centers around the world.

☐ I know data doesn't actually float in the sky — it lives in physical machines on Earth.

☐ I can trace what happens when I click a link, from my browser to the server and back.

☐ I know that TCP/IP, DNS, and HTTP are the invisible glue holding the Internet together.

☐☐ Section 2 – Browsers, Requests & DNS

☐ I can describe what happens when I type a URL in my browser.

☐ I understand that DNS is like the Internet's phone book — it turns domain names into IP addresses.

☐ I know that truehost.com and 104.21.12.145 are basically the same thing, just one's for humans and one's for machines.

☐ I can explain what an IP address is (and why IPv6 exists).

☐ I know how caching helps websites load faster.

☐ Section 3 – Servers & Hosting

☐ I know that a server is just a computer that serves other computers.

☐ I can describe the difference between shared hosting, VPS, dedicated servers, and cloud hosting.

☐ I understand that "serverless" doesn't mean "no servers" — it means someone else manages them.

☐ I know how data centers work — including power, cooling, and security systems.

☐ I realize "uptime" is not luck, it's engineering (and redundancy is the secret sauce).

☁☐ Section 4 – The Cloud in Real Life

☐ I can define the three types of cloud services: IaaS, PaaS, and SaaS.

☐ I understand what "regions" and "availability zones" are in cloud infrastructure.

☐ I know how CDNs make websites faster by serving cached copies closer to users.

☐ I realize cloud scaling happens automatically — adding more servers as demand grows.

☐ I can explain what edge computing is (and why it's making the Internet faster).

🔐 Section 5 – Security & Reliability

☐ I can spot the difference between HTTP and HTTPS (and know the padlock means encryption).

☐ I understand what a firewall does — and that it's not just software but a whole class of defenses.

☐ I know why multi-factor authentication (MFA) matters.

☐ I can explain what a backup is and why you should have three copies in two places.

☐ I know redundancy isn't optional — it's the reason the Internet doesn't collapse daily.

☐ I can describe what happens during a DDoS attack (and how CDNs and firewalls help stop it).

☼☐ Section 6 – Fun Geek Facts (Bonus Round)

☐ I know there are only 13 *root* DNS server clusters (replicated worldwide).

☐ I know the Internet uses fiber-optic cables that literally run under the ocean.

☐ I know that Google, Amazon, and Microsoft operate hundreds of data centers globally.

☐ I know "the cloud" is powered by clean energy more often than not.

☐ I can finally tell my friends that "the cloud is just someone else's computer."

💡 Section 7 – Quick Recap in 10 Sentences

1. The Internet started as ARPANET — a military project to connect computers that could survive an attack.
2. Every device online has an IP address, like a digital mailing label.
3. DNS turns names into numbers so browsers can find the right server.
4. When you click a link, your browser builds a request and sends it through routers and ISPs to a data center.
5. Servers inside that data center fetch or generate the requested page.
6. The page travels back to you through the same network in tiny packets.
7. Your browser reassembles it into the site you see.
8. Redundancy, load balancing, and backups make sure the site doesn't disappear.
9. The cloud makes this scalable — adding or removing servers automatically.
10. Every website, app, and cat video depends on this invisible collaboration happening billions of times per second.

🎁 Section 8 – Real-World Reflection

☐ I can confidently explain what "the cloud" means to a friend or coworker.

☐ I understand that uploading a file "to the cloud" means storing it in someone's data center.

☐ I know what a domain registrar and hosting provider actually do.

☐ I could (in theory) set up my own website from scratch, understanding every piece of the puzzle.

☐ I realize that the Internet isn't magic — it's engineering, cooperation, and a whole lot of blinking lights.

🏁 Congratulations — You're Now an Internet Insider

If you can check most of these boxes, give yourself a pat on the back. You've just demystified something that even many "techies" take for granted.

You now understand:

- Where your data goes,
- How websites stay alive,
- Why "the cloud" matters,
- And what's really happening when you click anything online.

You're no longer just a user — you're an informed participant in the greatest

3.10 📖 Glossary – "The Internet in Plain English"

(Because every good explorer needs a map — even in cyberspace.)

A

API (Application Programming Interface)
A set of rules that lets different software systems talk to each other.
Example: When your phone's weather app pulls data from a weather website, it's using an API.

App (Application)
Software designed to do specific tasks. Apps can live on your phone, laptop, or even run in your browser (like Gmail or Netflix).

ARPANET
The ancestor of today's Internet — a 1960s U.S. military project connecting research computers so they could communicate in case of attack.

B

Backup
A copy of your data stored separately so you can restore it if the original is lost, deleted, or corrupted.

Bandwidth
How much data can pass through a network in a given time — like the width of a highway for your Internet traffic.

Browser
The software (like Chrome, Safari, or Firefox) that you use to view websites. It translates code (HTML, CSS, JavaScript) into the pages you see.

Bug
A flaw in code that causes weird or unwanted behavior — basically, software's version of sneezing.

C

Cache
A temporary storage area that keeps copies of files or data for faster access later. Browsers cache images and pages so they load faster next time.

CDN (Content Delivery Network)
A global network of servers that store cached website content closer to users — making sites load faster no matter where you are.

Cloud Computing
Renting computing power, storage, or apps from someone else's servers instead of running them on your own machine.

Colocation
When a business places its own physical servers inside someone else's data center for power, cooling, and connectivity benefits.

Cookie
A small file saved on your device by a website to remember who you are — good for keeping you logged in, not so good when it tracks too much.

D

Data Center
A giant facility filled with servers, cooling systems, and cables — the physical home of "the cloud."

Database
An organized collection of data that can be easily searched and updated. Think of it as a digital filing cabinet.

DDOS (Distributed Denial of Service)
An attack where hackers flood a website with traffic, overwhelming it until it slows down or crashes.

DNS (Domain Name System)
The Internet's phonebook — it turns easy names like truehost.com into hard numbers like 104.21.12.145.

Domain Name
The human-friendly label for a website (like google.com or olitt.com).

Download
Copying data from the Internet onto your device. The opposite of **upload**.

E

Edge Computing
Placing servers physically closer to users to reduce latency and speed up real-time processing.

Email (Electronic Mail)
The OG of Internet communication — sending digital messages across networks using addresses like you@truehost.com.

Encryption
Scrambling information so only the intended recipient can read it — like locking your data in a digital safe.

F

Firewall
A system that monitors and filters incoming or outgoing network traffic based on security rules — your first line of digital defense.

FTP (File Transfer Protocol)
One of the oldest Internet protocols for transferring files between computers.

G

Gateway
A device or software that connects different networks — like a translator between two countries.

Gigabyte (GB)
Roughly one billion bytes of data. Enough to hold about 250 songs or one short HD movie.

H

Hosting
Renting space on a server so your website can live online.

HTML (HyperText Markup Language)
The code that structures a web page — text, links, and layout. CSS and JavaScript handle style and interactivity.

HTTP / HTTPS
The rules your browser and a web server use to talk to each other.
The "S" in HTTPS means it's secure (encrypted).

I

IaaS (Infrastructure as a Service)
Cloud model where you rent virtual servers and storage — you handle the software. (Example: AWS EC2.)

IP Address (Internet Protocol Address)
A unique number that identifies every device connected to the Internet.
Example: 192.0.2.44

ISP (Internet Service Provider)
The company that connects you to the Internet — like Comcast, Verizon, or Spectrum.

J–L

LAN (Local Area Network)
A small, local network (like your home Wi-Fi) connecting nearby devices.

Latency
The delay between sending a request and getting a response — the Internet's version of "lag."

Load Balancer
A system that distributes website traffic across multiple servers to prevent overloads.

M

MFA (Multi-Factor Authentication)
A security feature requiring two or more verification methods (password + code texted to your phone).

MX Record
A DNS record that tells the Internet where to deliver email for your domain.

N

Name Server
A DNS server that stores and answers queries about domain names and their IP addresses.

Network
A group of connected computers that share data and resources.

O–P

Packet
A small chunk of data sent across the Internet. Every web page, image, or video you load is broken into packets and reassembled later.

PaaS (Platform as a Service)
A cloud service where developers build apps without worrying about infrastructure. (Example: Heroku, Firebase.)

Ping
A test that checks how quickly one device can reach another across the Internet.

Protocol
A set of rules that define how computers communicate (like HTTP, FTP, or SMTP).

R–S

Redundancy
Having duplicate systems or backups in place to ensure uptime even when something fails.

Router
A device that directs Internet traffic between networks — the reason your Wi-Fi connects to the world.

SaaS (Software as a Service)
Cloud software you access through a browser — like Gmail, Zoom, or Canva.

Server
A computer that stores and delivers data to other computers (clients) over a network.

SSL / TLS (Secure Sockets Layer / Transport Layer Security)
The technology behind HTTPS — encrypts data between your browser and the server.

T–U

TCP/IP (Transmission Control Protocol / Internet Protocol)
The foundational rulebook of the Internet — how data is broken into packets, sent, and reassembled.

TTL (Time To Live)
A setting that controls how long DNS data is cached before being refreshed.

Upload
Sending data from your device to the Internet (like posting a photo or video).

URL (Uniform Resource Locator)
The complete web address you type in your browser — includes the protocol, domain, and sometimes page path.
Example: https://www.truehost.com/contact-us

V–Z

VPS (Virtual Private Server)
A virtual slice of a physical server that behaves like its own private machine.

VPN (Virtual Private Network)
Encrypts your Internet connection and routes it through another location — great for privacy and accessing region-locked sites.

Web Hosting
A service that keeps your website's files online and accessible.

WWW (World Wide Web)
The collection of websites and web pages accessible via browsers — technically, it's just one part of the Internet.

Bonus Term

"The Cloud"
A marketing-friendly way of saying, *"someone else's computer you rent time on."*
It's everywhere — powering everything from your phone backups to your favorite streaming shows.

✨ That's It!

You've reached the end of BEHIND THE CLICK: The Untold Story of the Internet
By now, you can officially:

- Read Internet diagrams without sweating,
- Explain DNS to your friends,
- And finally, roll your eyes (knowledgeably) when someone says, "It's in the cloud."

Made in United States
Cleveland, OH
04 November 2026

47645292R00078